173 Topics in Current Chemistry

Molecular Similarity I

Editor: K. Sen

With contributions by
N. L. Allan, E. Besalú, R. Carbó,
D. L. Cooper, J. Mestres, P. G. Mezey,
D. H.Rouvray, M. Solà

With 23 Figures and 24 Tables

Springer-Verlag
Berlin Heidelberg GmbH

This series presents critical reviews of the present position and future trends in modern chemical research. It is addressed to all research and industrial chemists who wish to keep abreast of advances in their subject.

As a rule, contributions are specially commissioned. The editors and publishers will, however, always be pleased to receive suggestions and supplementary information. Papers are accepted for "Topics in Current Chemistry" in English.

ISBN 978-3-662-14891-4   ISBN 978-3-540-49039-5 (eBook)
DOI 10.1007/978-3-540-49039-5

Library of Congress Catalog Card Number 74-644622

© Springer-Verlag Berlin Heidelberg 1995
Originally published by Springer-Verlag Berlin Heidelberg New York in 1995
Softcover reprint of the hardcover 1st edition 1995

Typesetting: Macmillan India Ltd., Bangalore-25
SPIN: 10474235   51/3020 - 5 4 3 2 1 0 - Printed on acid-free paper

Guest Editor

Prof. Dr. *Kali Das Sen*
University of Hyderabad
School of Chemistry
Central University P.O.
Hyderabad 500 134, India

Editorial Board

Attention
all "Topics in Current Chemistry" readers:

A file with the complete volume indexes Vols.22 (1972) through 172 (1994) in delimited ASCII format is available for downloading at no charge from the Springer EARN mailbox. Delimited ASCII format can be imported into most databanks.

The file has been compressed using the popular shareware program "PKZIP" (Trademark of PKware Inc., PKZIP is available from most BBS and shareware distributors).

This file is distributed without any expressed or implied warranty.

To receive this file send an e-mail message to:
SVSERV@VAX.NTP. SPRINGER.DE
The message must be:"GET/CHEMISTRY/TCC_CONT.ZIP".

SVSERV is an automatic data distribution system. It responds to your message. The following commands are available:

HELP	returns a detailed instruction set for the use of SVSERV
DIR (name)	returns a list of files available in the directory "name",
INDEX (name)	same as "DIR",
CD <name>	changes to directory "name",
SEND <filename>	invokes a message with the file "filename",
GET <filename>	same as "SEND".

For more information send a message to:
INTERNET:STUMPE@SPINT. COMPUSERVE.COM

Preface

In the past, similarity, or the lack of it, has been studied qualitatively in almost every discipline of knowledge. Recently, chemists have devoted considerable attention to devising several useful quantitative measures which have made the similarity concept of fundamental importance in molecular engineering. The present two volume monograph on 'molecular similarity' is an attempt to cover some of the main conceptual and computational development in this rapidly growing multidisciplinary area of research. It is hoped that several other quantum chemists (density-and wave-functional) will assist in refining the molecular similarity measures so as to make them uniformly applicable to small and large molecules and, equally importantly, contribute to defining other new measures.

I take this opportunity to place on record my deep sense of gratitude to Professor C. N. R. Rao, F. R. S., for introducing me to the fascinating world of monograph editing.

Kali Das Sen

Table of Contents

Similarity in Chemistry: Past, Present and Future

Dennis H. Rouvray

Department of Chemistry, University of Georgia, Athens, Georgia 30602-2556, USA

Table of Contents

This opening article on similarity starts with a brief historical introduction to the subject and then considers the role played by similarity in the sciences at the present time. It is shown that all scientific concepts and classifications have a basis in similarity. It is also pointed out that similarity assessments are always to some extent arbitrary and so the concept can be defined only in relative terms. A detailed analysis then follows of the various kinds of similarity that may be used in chemical applications: analogy, complementarity, equivalence relations, scaling and self-similarity. For each of these kinds instances of current applications are given. Analogy can be further subdivided into five varieties, two of which – functional and inductive analogy – are shown to be of fundamental

Topics in Current Chemistry, Vol. 173
© Springer-Verlag Berlin Heidelberg 1995

importance in molecular design. In discussing complementarity, we address the mapping of biological receptors and various measures for molecular shape are outlined. After completing our survey of the differing kinds of similarity, we close by taking a peek into the future and assess the possible roles that may be played by fuzzy logic and the use of neural networks in similarity studies.

1 Introduction

Similarity is an idea whose time has come. Why it should have come now is not in doubt. The surge of interest in similarity witnessed in recent years has been driven by one primary goal: exploitation of the great practical utility of this concept in the chemical domain. Its current popularity is likely to be long-lasting, for the emergence of similarity on the scene can in no way be viewed as a fashionable trend that will soon be dissipated. Similarity is actually a very old concept that has been with us from the earliest times. It has certainly played a vital role in the historical development of chemistry. Indeed, it is no exaggeration to assert that, without the extensive application of this concept in the past, chemistry could not have developed at all [1]. To understand why this is so, we shall begin by taking a brief look at the origins of this particular science – and, concurrently, all modern sciences. Scientific thinking first evolved in ancient Greece, roughly during the period 500–300 B.C. Hence it is to the genius of ancient Greek thought that we are indebted for our present-day sciences. Greece bequeathed to us the concepts of objectivity and deductive reasoning on which all sciences are based [2]. It is also not surprising that we encounter the earliest uses of the notion of similarity in a scientific setting in the works of certain ancient Greek philosophers [3]. To lay the necessary foundations for our subsequent discourse, we shall explore initially some of the original contributions to our subject made by Greek thinkers almost 2500 years ago.

The first Greek contribution is a variety of similarity called analogy. The term is actually derived from the Greek *analogia* and originally referred to proportionality in the sense that the sides of similar geometric figures, e.g. triangles, are analogous because they are proportional in length. Many Greek philosophers, starting with Empedocles (492–432 B.C.), made extensive use of the concept. Over a period of time the concept gradually changed in meaning until it came to refer ultimately to a direct comparison between two sets of relationships characterizing two different systems. Plato (427–347 B.C.) was so enamoured of the concept that he declared analogy to be "the most beautiful form of argument" [4]. Even today, analogy is still widely viewed as an exceptionally powerful method of reasoning [5]. There is good reason to make such an assertion, for analogy alters the context of our perception of reality and so makes it possible to approach seemingly intractable problems in creative new ways. In the case of the evolution of chemistry, analogies have played an indispensable role. Some of the many analogies employed in various chemical contexts over the past 2500 years are presented in Table 1. The type of analogy

Table 1. A listing of some of the analogies used to interpret chemical phenomena over the past 2500 years

Proponent	Year	Description of proposed analogy
Democritus	ca. 420 B.C.	Atoms are like tiny particles that differ in both shape and size
Boyle	1660	Gases consist of particles that behave like miniature springs
Newton	1718	Atoms are like hard, impenetrable, movable spheres
Bernoulli	1738	Gases behave like billiard balls in a box
Black	1803	The flow of heat through a system resembles the flow of water
Dalton	1804	Molecules are analogous to hard spheres connected together by sticks
Davy	1806	Electrolytes behave like hard, charged spheres
Kekulé	1865	The structure of benzene is like that of a snake biting its tail
Baeyer	1885	Molecules are analogous to hard spheres connected together by springs which may contain strain energy
van't Hoff	1888	Dilute solutions behave like gases
Fischer	1894	A drug at a biological receptor behaves like a key turning a lock
Thomson	1899	Electrons collectively behave like clouds
Nagaoka	1904	Atoms resemble planar solar systems
Rutherford	1911	Atoms resemble three-dimensional solar systems
Bohr	1939	The atomic nucleus behaves like a drop of liquid
Rouse	1953	Long polymer molecules behave like beads on a spring
Pearson	1963	Hard likes hard and soft likes soft; HSAB principle
Kroto et al.	1985	The buckminsterfullerrence molecule C_{60} behaves like a miniature soccer ball
Lutz and Schmidt	1992	Chemical cluster molecules behave like the atomic nucleus

introduced by the Greeks – now generally described as proportional analogy – has been largely superseded by other varieties of analogy, notably functional and inductive analogies, which we address later on in our exposition.

The second Greek legacy is the notion of the atom as a fundamental particle of matter. The Greeks speculated extensively about the nature of atoms and reached a number of surprisingly accurate conclusions. The word atom was first coined by Leucippus (ca. 470–420 B.C.) to signify a tiny speck of matter. He maintained that atoms are so small that they are forever invisible, though they continually move around in empty space. He also believed that atoms exists in a variety of different types and that the differing types would differ in shape. Whenever atoms clustered together, there would also be differences in their relative positions and arrangements within such clusters. Many of these ideas were also echoed by Democritus (ca. 460–370 B.C.), a contemporary of Leucippus. Democritus asserted that atoms were infinitely hard and so could never break. He also taught that atoms had weight and differed in size. Spherical atoms were deemed by him to be the smallest of all the possible shapes. When different kinds of atoms became entangled to form clusters, each atom would keep its individual identity as a certain amount of space would always separate the atoms. The differences observed in the manifold types of bulk matter were due ultimately to variations in both the shapes of the atoms and the closeness and arrangement of their packing. It has been claimed that such ideas adumbrate the modern concept of isomerism [6]. The notion that substances could be

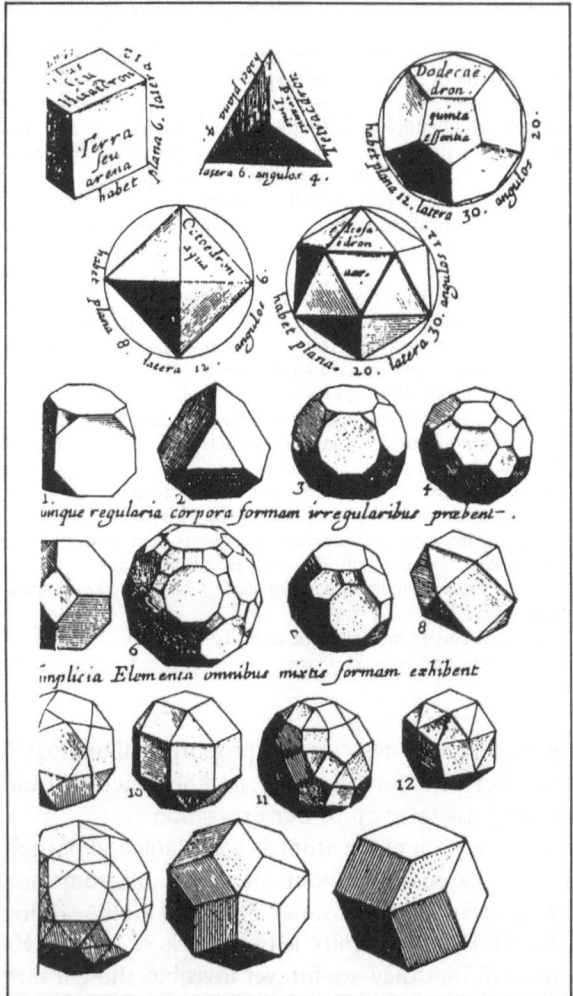

Fig. 1. The scheme of Davisson for the analysis of substances in terms of the five Platonic polyhedra. Taken from Philosophia Pyrotechnica, part 3, Paris, 1642

analyzed in terms of solid figures of various shapes persisted well into the seventeenth century, as will be apparent from Fig. 1.

The third and final Greek contribution that we shall mention here is the notion of classification. The ancient Greeks were classifiers *par excellence* and classified everything from atoms to animals. In fact it was Aristotle who first attempted a classification of the various species of fauna [7]. The process of classification always involves the idea of similarity; in this process objects that have at least some features in common, i.e. some similarity, are grouped together to form classes or categories. The process was explored in some detail in a book on the subject of categories by Aristotle, who made the prescient statement that "those things are contraries which, within the same class, are separated by the greatest possible distance" [8]. The significance of this remark will become clearer when we discuss the topic of molecular design in Sect. 4. Modern researchers [9, 10] are agreed about the key role played by similarity in the process of category formation; however, as Oden and Lopes [11] have warned "although similarity must function at some level in the induction of concepts, the induced categories are not 'held together' subjectively by the undifferentiated 'force' of similarity, but rather by structural principles." In other words, the idea of similarity is important in both the induction of concepts and in the formation of categories, but it is not the only factor that needs to be taken into consideration. This is a theme we explore more fully in the following section.

2 Concepts and Categories

After our brief excursion into the classical arena to inspect some of the earliest entries in the similarity cavalcade, we now leapfrog the centuries to return foursquare to the present time. For the concept of similarity to be applicable in a modern scientific setting, it is necessary that it have considerable flexibility. It must be sufficiently broadly defined that it can cope, among other things, with the classification of diverse information into well-defined categories, the expression of relationships in a wide variety of conceptual spaces, the discernment of patterns in complex data sets, and the formulation of scientific hypotheses and laws. The first step involves envisaging some part of reality that has been isolated for separate study; such a part is usually referred to as a domain of knowledge [12]. Typically, a domain of knowledge consists of a set of concepts together with a family of mathematical relations defined on this set. As there is no obvious way in which such a domain can be excised from the rest of the world, partitionings of this sort are always arbitrary. In fact, the situation has been neatly summed up by Bateson [13] who said that "the division of the perceived universe into parts and wholes is convenient and may be necessary, but no necessity determines how it shall be done."

The structuring of reality is accomplished in the brain by matching up incoming data with information already stored in the memory. Objects and events that are perceived are thereby grouped into concepts and categories, though such groupings are tentative in that they may be revised in the light of new experiences. The brain normally operates with soft-edged or fuzzy groupings rather than the more rigorously defined categories characteristic of the physical sciences [14]. A concept is now defined as a mental construct that includes everything that is associated with this construct at some given point in time. Such a statement implies that (i) the definition itself is vague; (ii) a concept is more than a mere idea; and (iii) concepts are unstable and so can change with respect to time. It has been postulated [10] that the process of concept formation involves first the perception of some apparent similarity in a set of observations and then the embedding of these observations in a web of relations of increasing levels of abstraction. Consider the concept of the acid [15]. In the ancient world, the Latin term *acidus* applied to anything that tasted sour such as vinegar. Mineral acids such as nitric and sulfuric acids were first prepared by medieval alchemists. In the 1920s Brønsted and Lowry defined acids as substances that could donate protons, thereby extending the acids to include substances such as ammonia and water. Lewis went further and redefined acids as electron pair acceptors and so introduced species such as boron trifluoride into the fold. The more recent addition of a plethora of amino acids, polybasic acids, and superacids has brought the number of acids into the thousands. Inclusion of all these varieties of acids has changed the concept dramatically; concepts usually change when they are embedded in some new contextual framework.

Because categories do not differ from concepts in any fundamental way, it is not surprising that both are equally difficult to define. Each is a construct formed during the mental process of separating off a part of reality from the rest of the world and each tends to change with time. A definition by Medin [10] that a category is a partitioning of the world to which some set of assertions may apply at some point in time is vague and also reveals why the terms concept and category are often used interchangeably – at least in the psychological literature. Medin [10] has also stated that "it is tempting to think of categories as existing in the world and of concepts as corresponding to mental representations of them, but this analysis is misleading." The reason for this is that individuals tend to impose structure on the world rather than discover it. In spite of this tendency, however, categories generally show a remarkable consistency across cultural and language barriers [1]. Typically, membership of a category is decided by reference to sets of (i) rules; (ii) prototypes; or (iii) exemplars [16]. Partitioning is carried out initially on the basis of the similarity of the objects under consideration, though the categories formed may eventually be restructured as deeper insights are gained. The addition or deletion of category members can have a decisive impact on both the nature of the category or on conclusions reached when making deductive inferences about the physical world.

Although there is no such things as a permanent category, it should be mentioned that there is great variability in the stability of categories with respect to time. Categories that are quasi-stable are those for which it is possible to specify some defining rule that exclusively identifies all its members; such categories are said to be well-defined [16]. Examples of well-defined categories in the realm of chemistry include most of the groups of compounds that have been designated with a specific name, e.g. the alkanes, the proteins, or the organometallics. Those categories for which the above does not apply are termed fuzzy categories [16]. Such categories are also encountered in chemistry, for instance, the category of aromatic molecules or the category of strong acids. The use of fuzzy set theory [17] or fuzzy logic [18] becomes appropriate for situations in which category membership is a matter of degree rather than a certainty or in cases where the boundaries of adjacent categories overlap. Fuzzy sets were introduced [19] in 1965 for the express purpose of characterizing systems with such inherent uncertainties about their makeup. The basic idea of a fuzzy set is that each member is neither definitely in nor out, but that it belongs to some extent. The assignment of the extent of belonging is achieved by associating with each member a function that is a real number on the line segment [0, 1]. This number is not to be interpreted as the probability that the member belongs but rather as a measure of how true it is that that member belongs to the set.

3 Manifestations of Similarity

Although we have so far made frequent mention of the concept of similarity, no attempt has been made to define it in a rigorous sense. There is good reason for deferring discussion on the nature of similarity until the preliminaries have been dealt with: the concept is a very elusive one. To try to get to grips with it, we shall first consider some relevant work in the psychological domain. There is substantial evidence to suggest that, in general, similarity calls are colored by emotional considerations, that is to say by the way in which we react emotionally to the objects being compared. After extensive experimentation, Tversky [20] concluded that similarity judgments were made on the basis of two quite different criteria, namely (i) intensive and (ii) diagnostic criteria. Intensive criteria are those that relate to nonsubjective factors, such as the intensity of the signal-to-noise ratio, the clarity of the vision, or the vividness of the observed objects. Diagnostic criteria on the other hand refer to our subjective response to the objects, such as their significance or salience to us as individuals. Because the human brain normally adopts both of these criteria in making similarity calls, a subjective component is usually present and, as a result, our calls have some degree of arbitrariness associated with them. However, even when the subjective component is minimal or removed altogether, some arbitrariness will remain because any measure used to estimate similarity will necessarily be arbitrarily selected.

In the scientific context, the attempt is made to reduce the importance of diagnostic criteria to an absolute minimum. Fortunately, this is comparatively easy to do when dealing with chemical entities such as structural formulas, for their emotive impact is clearly very small. Moreover, similarity measures can be adopted that are algorithmic in nature, i.e. measures that involve no actual similarity calls. Tversky [20] has also investigated an asymmetry in similarity calls that can arise when the sequence of the presentation of objects is altered. This problem is typically encountered when one of the objects being compared is regarded as a prototype. In comparison of circles and ellipses, for instance, experimental observers consistently reported that ellipses were more similar to circles than circles to ellipses. The rule here appears to be that a variant will always be judged more similar to a prototype than vice versa. Another psychological study, by Krueger [21], has revealed that the judging of two patterns to be the same was faster than pronouncing them to be different. This is somewhat surprising as judgments of sameness require exhaustive processing, unlike judgments of difference for which only one difference will suffice. This finding is explained by assuming that the internal noise in the brain is more likely to make identical patterns seem different than conversely [21]. To compensate for the noise, the brain rechecks all different calls and so spends more time on them. The effect can be experienced by comparing the two pairs of structures in Fig. 2: the identical pair is the more quickly recognized of the two pairs.

Because similarity assessments depend on the structure and function of the human brain, some aspects of assessing similarity are universal whereas others are idiosyncratic in that they pertain to the individual observer. The universal part of such assessment involves stimuli entering the eyes, passing through the eyes, and ending up in the cerebral cortex of the brain. Upon arrival, the stimuli trigger the firing of millions of neurons throughout the cortex in a pattern that has been likened to that of physical chaos [22]. The chaotic aspect is evident in the tendency of large numbers of neurons to switch suddenly from one firing

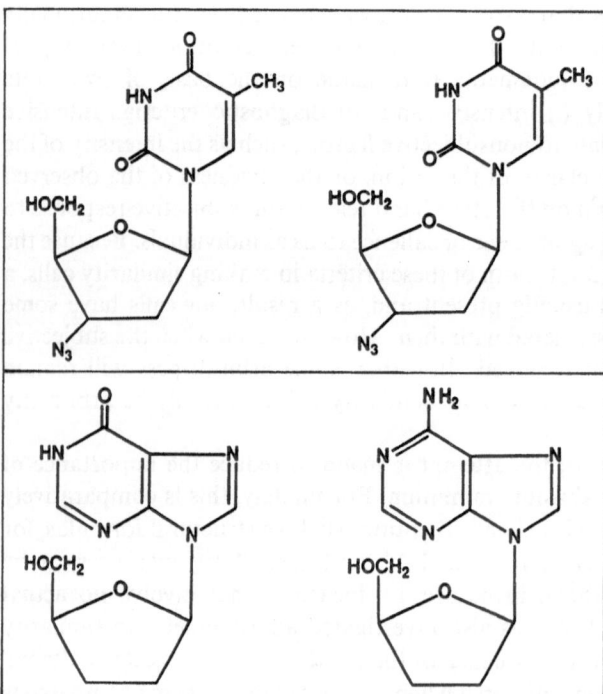

Fig. 2. Two pairs of molecular structures, one pair showing identical structures and the other pair different structures. Surprisingly, it takes longer to identify the differing pair than the identical pair. The structures represent drug molecules known to be active against the HIV virus that causes AIDS

pattern to some other pattern in response to tiny stimulatory inputs. This capability is key, for it is what makes perception possible in the first place. The interaction between the neural activity generated by the stimuli and the ongoing level of activity in the cortex constitutes the output of the brain. Similarity judgments typically require a multidimensional evaluation of several different features of the objects being compared, including their color, form, movement, and stereopsis [23]. Such features are referred to as the dimensions of the similarity. It is the need to assess simultaneously several dimensions that creates the difficulty in making similarity calls, and is also the reason that there can, in general, be no such thing as "pure similarity" [24]. The idiosyncratic aspect to similarity assessment arises because each human cortex has a differing background noise level and also each individual has differing emotive responses to the various dimensions of similarity that are to be judged.

The end result of the various difficulties encountered in the assessment of similarity is that it is not possible to define any absolute measures of similarity. All similarity determinations are thus relative and the best that can be done is to define relative measures of similarity. This is not necessarily a major drawback, for even relative similarities can be extremely valuable to chemists. In fact, relative similarities have found widespread expression in many different chemical contexts. In the process of accommodating the concept to the manifold roles it has been required to play, the concept has become ever more elaborated. Currently, we have at our disposal a wide range of varieties and subvarieties of similarity; several of the varieties are listed in Table 2. One of these varieties – analogy – has already been mentioned in our introduction; this particular variety

Table 2. A listing of several of the varieties of similarity that can be exploited in the scientific context

Similarity Relationship	Author(s)	Date
Equivalence of acids and bases	Richter	1793
Similarity of atoms of an element	Dalton	1808
Isomorphism of crystals	Mitscherlich	1818
Isomerism of molecules	Berzelius	1830
Similarity of organics and inorganics	Dumas	1839
Allotropy of chemical elements	Berzelius	1840
Homology in hydrocarbons	Gerhardt	1845
Classification of molecular types	Hofmann	1849
Classification of reaction types	Kekulé	1858
Periodic classification of elements	Mendeleev	1869
Stereosiomerism of molecules	van't Hoff, Le Bel	1874
Isotopy of atoms	Soddy	1913
Isosterism of molecules	Langmuir	1919
Electronic classification of atoms	Pauli	1925
Classification of rigid molecule sates	Herzberg	1945
Homology in biomacromolecules	Fox, Homeyer	1955
Classification of nonrigid molecule states	Longuet-Higgins	1963
Classification of biological receptors	Ash, Schild	1966
Periodic classification of benzenoids	Dias	1985
Classification of molecular shape	Mezey	1985

has been divided into at least five subvarieties [5]. Some of the other varieties that we shall be considering include the equivalence relationship, partial and hierarchical ordering of sets, complementarity, scaling, and self-similarity. For each of these varieties we shall give a brief description of the type of similarity involved, followed by reference to some of the principal applications in the chemical domain. From our presentation, it will be clear that the type of similarity adopted in any specific instance must be carefully selected and matched against the problem in hand, for the inappropriate use of similarity can do more harm than good. By giving some general guidelines in our subsequent discussion, we hope that the reader will thereby be enabled to make a judicious selection whenever the need arises.

4 The Role of Analogy

As indicated in our introduction, analogy is not only a very old concept but also one that is exceptionally powerful. Some caution in its use is therefore advisable. Analogy involves a kind of similarity in which a more complex system is described and interpreted in terms of a simpler and supposedly better understood system. Analogy thus involves modeling; an atom might, for instance, be modeled as a miniature solar system because the latter is thought to be well understood. Such modeling necessarily implies the existence of at least some correspondence between the two systems, and so it becomes possible to map certain parts of each system on to one another. Such a mapping typically entails the mapping of some of the attributes of the first system on to some of the attributes of the second system in a one-to-one mapping. The attributes mapped may be either components of the two systems or relationships that exist between those components. To refer back to the analogy of the atom and the solar system again, the nucleus would be mapped onto the sun and the electrons onto the planets. However, the electrostatic attraction between the nucleus and an electron could equally well be mapped onto the gravitational force between the sun and a planet. Of course, this rather simplistic view of the atom has now been superseded; the advent of quantum mechanics consigned this particular model of the atom to history [6]. The analogy, however, is still an instructive one if for no other reason than it reminds us of the great fruitfulness of many of the analogies that have been drawn in the past.

At this point we move on to a more mathematical characterization of analogy. The concept of analogy has been subdivided into at least five varieties: (i) attributive; (ii) functional; (iii) inductive; (iv) proportional; and (v) structural analogies. Of these, we shall discuss only functional and inductive analogies; all five have, however, been discussed in considerable detail elsewhere by the present author [5]. A functional analogy is defined either for two objects that have the same function or for two systems in which certain of the components play the same role. In the first case we have a global mapping from one object to the other; in the second case we have mappings between relationships. An instance of the first case is afforded by samples of mercuric fulminate and trinitrotoluene: though different chemically, both can be used as explosives. An instance of the second case is provided by the rifamycins and the streptovaricins, two different drugs that both display antibacterial activity because they have structures that can inhibit RNA synthesis in pathogens [25]. The inductive analogy is defined for some set of objects A, B, C, etc. all of which possess some attributes, such as P and Q. If the objects B and C are subsequently shown to exhibit the attribute R, then by inductive analogy we may reason that A will also likely exhibit attribute R. An instance of such reasoning is afforded by the set of penicillin molecules that are known to be bioactive and also nontoxic. By inductive analogy, any new penicillin molecule that is synthesized and found to be bioactive will also probably be nontoxic. This latter variety of analogy takes us one step further than the rather general and nonspecific reasoning furnished by the functional analogy.

To illustrate this latter point, we show in Fig. 3 two systems that are comprised of the components A_1 through G_1, and A_2 through G_2. Some of the components in system 1 can be mapped onto components in system 2 in a one-to-one mapping. Either objects or relationships may be mapped in this way. Those components or relationships that can be mapped are said to constitute the positive analogy, those that cannot be so mapped the negative analogy. The inductive analogy comes into play when the behavioral characteristics of the two systems are compared. For systems with a large positive analogy, the reasoning is that if system 1 exhibits some kind of behavior α, and the equivalent type of behavior in system 2 is β, then α and β will be similar. Such reasoning is known as argument from the Principle of Analogy [5]. Examples of application of this Principle include (i) the use of the concept of homology which states that successive members in homologous series will exhibit similar behavior [26]; (ii) the more general principle – which was first enunciated in ancient Greece by Hippocrates – that like substances associate with each other because they have similar characteristics [27]; and (iii) the many variants of the Principle of Least

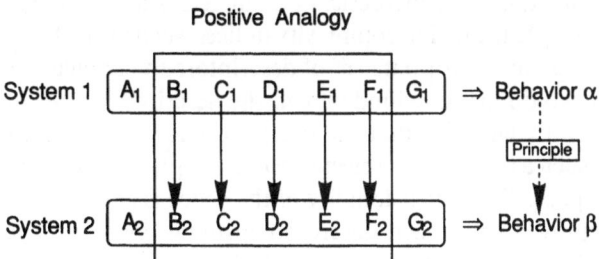

Fig. 3. A diagrammatic depiction of a family of one-to-one mappings from one system to another system to illustrate the nature of the analogy

Action [28], such as Le Chatelier's Principle [29], which states that whenever a chemical system undergoes some change, it will always do so in a way that leaves the reactants and products as similar as possible.

One major application of analogy of special interest to us here has been its use in classifying and predicting the behavior of both atomic and molecular species. Such uses are very old, one of the earliest being that of Döbereiner who in 1829 recognized the existence of analogies between various chemical elements [30]. He showed that certain elements form groups of three or triads in which the middle element has properties that are the mean of the other two. Examples of triads he mentioned are calcium-strontium-barium and chlorine-bromine-iodine. These ideas generally became elaborated over the next forty years until by 1869 the full Periodic Table of the elements had been developed [31, 32]. The logic used in more recent times has been extended even further in that many Periodic Tables based on compounds have also now appeared. Examples of such tables include those for diatomic species [33, 34], triatomic species [35], and a wide variety of organic species ranging from polycyclic aromatic hydrocarbons to fullerene clusters [36, 37]. In the design of molecules for specific applications, use is made of inductive analogy. This means that the Principle of Analogy [5] is invoked as mentioned above. In such cases, the principle informs us that molecules with similar structures will also likely exhibit similar biological or physiological activity. This reasoning has led to the search for measures of similarity in molecular structures, of which many have been defined. We now consider the theoretical basis for these measures.

It is the great merit of the Principle of Analogy that it enables us to cluster together like molecules and then to make predictions as to how such molecules will behave. The use of similarity-based methods in the area of molecular design, which has excited considerable interest over the past decade [38], involves two fundamental problems. These are finding an appropriate representation for individual chemical species and devising a means of clustering those species that are most similar. Chemical species can be represented in manifold different ways, but what is particularly needed here is a way that yields a numerical parameter. Such parameters can be obtained from (i) physicochemical or thermodynamic properties such as the octanol-water partition coefficient or the heat of formation; (ii) topological indices [39], such as the Wiener index [40] or the molecular connectivity indices [41]; (iii) quantum-theoretical parameters, such as charge densities or shape descriptors [42]; and (iv) complexity indices, such as the Bertz index [43]. In the selection of an appropriate set of descriptors to characterize species, it is important to avoid having redundancy in the descriptors, a result that can be simply achieved by ensuring that the descriptors form a mutually orthogonal set [44]. Once such a set has been selected, the descriptors are used to form a multidimensional space. The dimensionality of the space is equal to the number of descriptors deployed, with each descriptor forming one axis. The chemical species under study are now plotted in this multidimensional space, and the next problem becomes one of determining how far apart pairs of points are in this space.

Clearly, the closeness of two points, referred to as their proximity [45], affords a measure of how similar the species are in the multidimensional space. Of the many methods developed to estimate proximities, most have adopted measures that satisfy the three criteria pertaining to metrics, namely they are non-negative distance functions that satisfy the conditions (i) $d(x, y) \geq 0$; (ii) $d(x, y) = d(y, x)$; and (iii) $d(x, z) \leq d(x, y) + d(y, z)$, where the distance function $d(a, b)$ represents the distance between the points a and b. The actual distance are often defined in terms of a Minkowski metric, i.e.

$$d(a, b) = \sum_{i=1}^{m} [|x_i - y_i|^r]^{1/r},$$

where m is the dimension of the space and r typically assumes the values 1 or 2. When $r = 1$, the so-called city block metric is obtained, and when $r = 2$ the very common Euclidean metric (which simply generalizes Pythagoras' Theorem) results. These differing kinds of distance measure are illustrated in Fig. 4. The distances can be employed as similarity measures, provided it is remembered that a large distance will represent a low similarity. It is thus more accurate to say that the distances represent measures of dissimilarity. If complete similarity (identity) is represented as unity and total dissimilarity by zero, we may then write:

Similarity $= 1 - d(a, b) = 1 - $ Dissimilarity.

The notion of structural similarity, and the measures that have been devised to assess it, have become of major importance over the past decade. In fact, the whole field of quantitative structure-activity relationships (QSAR) has been a burgeoning one during this period. So it is hardly surprising that a large number of different methods of comparing molecules to estimate their relative similarities have emerged [46]. A systematic study of several measures of intermolecular structural similarity revealed [47] that only minor differences exist between the various measures, and that consequently all of them could be regraded as

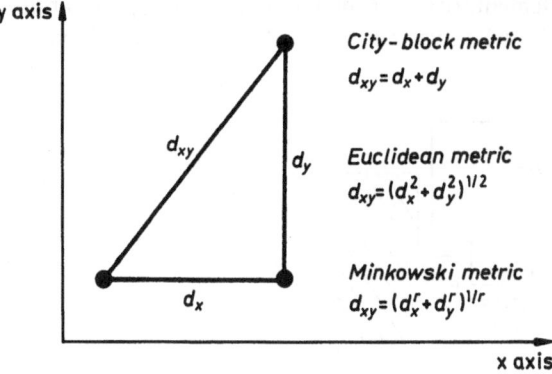

Fig. 4. An illustration of three different distance measures used in the estimation of intermolecular similarity in a two-dimensional representation

yielding satisfactory results. In such a situation, the method of choice often becomes the one involving the minimum of computation. But, whatever measure is ultimately adopted, it must be remembered that these are relative and not absolute measures. Moreover, all molecular design methods that exploit similarity rest ultimately on the Principle of Analogy. This can have obvious disadvantages if the wrong analogies are made. As Gould [48] has warned, "analogous similarities are particularly difficult to evaluate because they may be meaningful or meaningless depending on context." In the case of inductive analogy, resort is made to an essentially psychological process and the underlying reasoning employed cannot be justified on logical grounds. To obviate the potentially negative impact that reasoning by analogy could have, rules are currently being developed to optimize analogy making [49], and constraints that can be imposed on the process are now being explored [50].

5 Chemical Complementarity

The concept of complementarity, which is used in both the theoretical and experimental spheres of chemistry, refers to the fitting together of two or more parts of some system or idea to form a composite whole. The concept, which is illustrated in Fig. 5, is of considerable importance in current chemical thinking. In the theoretical arena, the concept has been extensively exploited in the description of the constituent particles of atoms, especially the electrons. The simultaneous particle-like and wave-like nature of such particles is thought to reflect a fundamental duality that is expressed as a complementarity in their behavior [51]. Of course, all objects have an associated wave function, though the wave nature of objects usually becomes progressively less important the larger the objects are. Thus, in determining the behavior of relatively large objects, the laws of classical physics tend to represent a very good approximation [52]. The concept of complementarity was first put forward in the atomic structure context by Bohr [53], who averred that, in discussing the behavior of

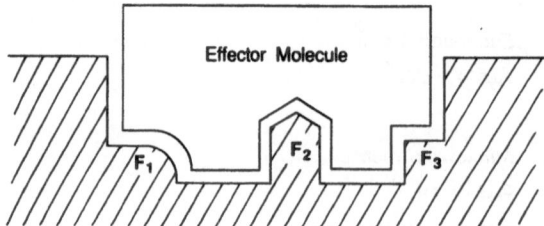

Fig. 5. A schematic representation of the fit of some receptor molecule at a biological receptor showing the complementarity of the interacting surfaces. Note that the functionalities F_1, F_2, and F_3 of the receptor are matched by the shape of the effector molecule

electrons, we meet with "a complementarity of the possibilities of definition quite analogous to that ... in connexion with the properties of light and free material particles" [53]. Although Bohr's ideas have been thoroughly analyzed in recent years, and even severely criticized for their vagueness and imprecision [54], it is clear that Bohr's basic understanding of complementarity was one involving a mutually or jointly completing set of concepts [54]. This is the sense in which the term is still widely used today.

In the experimental domain, one of the earliest uses of the concept was in a rudimentary description of the action of a drug molecule at a biological receptor. In 1894 Fischer [55] put forward the idea that a drug interacts with its receptor just as a key fits into and then opens a lock. This is an idea that has been substantially refined in recent times, though it is still an apt analogy in many ways. The interaction of an effector molecule at a biological receptor is now viewed as a very elaborate kind of complementarity which requires that biological systems be described in both structural and informational terms for their complete characterization [56]. Indeed, Greenspan [57] has gone as far as developing a comprehensive itemization of the various factors involved. These are the factors of (i) dimensionality, i.e. the number of different variables that can be used to characterize effector molecules; (ii) cooperativity, i.e. modification of structural features in the presence of other such features;. (iii) hierarchy, i.e. variability in the degree to which constituents involved in bioactivity actually contribute to that activity; and (iv) level of resolution, for a description that adequately characterizes an interaction at one level, e.g. the atomic level, may not be appropriate at some higher level, e.g. the amino acid residue level. Thus, Greenspan [57] has underscored the need for a complete description of interacting species that involves information on the relative energetic contributions of both atoms and amino acids, i.e. a hierarchical listing of the structural elements as well as the corresponding spatial coordinates.

At present it is possible to synthesize structures containing cavities or clefts that can be matched to virtually any small ion or molecule [58]. In biochemical or biological contexts, however, our expertise and knowledge are far more restricted. Although information about potential receptor sites on peptides and nucleotides has increased dramatically in recent years, it is still fair to claim that at present many biological receptors remain *terra incognita*. Because the recognition of substrates by such receptors is of crucial importance – many life processes depend on it – a multipronged approach to the mapping of receptor sites is now under way. The advent of molecular graphics devices that enable complex molecular structures to be manipulated on a computer video screen has assisted greatly in the process. With the aid of such devices it is currently possible to evaluate the docking of effector molecules at receptor sites. For instances, Meng et al. [59] studied docking in terms of molecular mechanics interactional energies to determine which molecular orientations were preferred, whereas DesJarlais et al. [60] used a docking algorithm to investigate goodness of fit for molecular sets at receptors with a view to discovering new lead structures in the drug design process. A comprehensive account of the use of three-dimensional structural formulas in such studies, together with relevant measures of structural similarity, has recently been published by Willett [61].

Ultimately, as our understanding of receptors increases, it will probably prove possible to classify them into various types. Some tentative steps in this direction have already been taken by Kenakin [62], who stressed that any such classification will need to be based on molecular constants of interaction that reflect the characteristics of both the drugs and the receptors. The elucidation of major receptor structures and the various kinds of interaction that take place at receptors will almost certainly lead to major advances in fields as diverse as pharmacy, medicine, chemical analysis, chromatography, catalysis, and materials science [63]. Up until very recent times the matching of receptor and docking molecular species has been largely guided by the shape of the docking species [64]. It thus seems appropriate here to say something about measures used to characterize shape and, in particular, the similarity of different shapes [42]. In general it may be said that shape is like similarity in that no absolute measure of shape can be defined. Moreover, the quantification of shape – even in relative terms – has proven to be an elusive task. Some of the earliest shape descriptors, based on the use of some reference structure, have been reviewed by Motoc [65]. The deployment of global descriptors, that is dimensionless numbers independent of the size of the object, has been discussed by Cano and Martínez-Ripoll [66].

At the molecular level, shape is now realized to be one of the most fundamental concepts of chemistry, even though it may be difficult to quantify [67]. In 1980 the first quantum-mechanical measure of shape similarity for molecules was put forward by Carbó et al. [68]. This measure, which was intended to be of special use in molecular design studies, proved to be of seminal influence. For any two molecules, M and N, the similarity, S_{MN}, was defined as the ratio

$$S_{MN} = \frac{\int \rho_M \rho_N \, dV}{[\int \rho_M{}^2 \, dV]^{1/2} \, [\int \rho_N{}^2 \, dV]^{1/2}},$$

where ρ_M and ρ_N are the electron densities for the molecules M and N respectively, and these densities are integrated over all space. In this expression the numerator affords a measure of the charge density overlap when the molecules are superimposed whereas the denominator may be regarded as a normalizing factor that produces values for S_{MN} that lie between zero and one. Since this measure was suggested it has been recognized that it suffers from certain shortcomings, chief among which is the fact that the S_{MN} values depend on the mutual orientation of the two molecules being compared. Not surprisingly, this expression has spawned a number of other formulas that allegedly represent improved versions of the initial attempt to quantify molecular similarity.

For instance, Hodgkin and Richards [69] proposed a modified version of the above expression that focused on the valence electron density in molecules rather than on the total electron density. It was believed that valence electron density would make similarity comparisons less arbitrary in that it could be employed to characterize (i) both the shape and the magnitude of the charge distribution in chemical species; (ii) the electrostatic potential surrounding a species (which affords a better discriminator for bioactivity); and (iii) the electrostatic field generated by species (which is of importance in dipolar interactions). However, this new formula still suffered from the drawback that the S_{MN} values depended on the mutual orientation and the distance separating the two molecules under comparison. This fact led Ponec [70, 71] to try out several different methods of optimizing the relative molecular positionings so as to maximize the S_{MN} values. His successes have led him to apply his methods to a detailed study of the similarity of reactants and products in pericyclic reactions [72]. Cooper and Allan [73] addressed the same problem concerning the S_{MN} values and came up with the idea of using momentum-space electronic densities rather than the more common position-space densities. This approach had a distinct advantage in that it circumvented several of the difficulties encountered in the earlier methods and also had the bonus of making possible the study of large series of molecules with the prospect of establishing structure-activity relationships [74].

The issue of molecular shape is likely to remain a controversial one. Indeed, even the existence of such a concept has been disputed [75]. Amann [76] has suggested that the concept of molecular shape is highly dubious because the shape interacts holistically with other molecular quantities and so cannot be defined in an unambiguous way. This reasoning, however, had not deterred a number of authors involved in similarity studies from tackling the problem of shape at the molecular level. Arteca and Mezey [77], for instance, have explored topological descriptions of the van der Waals surfaces of molecules. Their prescription is first to decompose the surface into spherical domains, i.e. two-manifolds, and then compute the homology groups for the topological entities into which the surface has been analyzed. The approach has the advantage that the final result is algebraic and so can be stored in a computer. Mezey [42] has pointed out that this approach is a powerful one as long as the molecular model used is defined in terms of classical mechanical analogies, though, in a rigorous sense, only certain of the essential and common features of the whole family of possible geometric models can be viewed as reflecting physical reality. Such considerations have led to the use of fuzzy sets [17] in the definition of shape similarity measures [78] and to the development of a so-called quantitative shape-activity approach (QShAR) to molecular design [79]. Other comparatively recent work that has explored ways of characterizing molecular shape includes the use of Fourier descriptors by Leicester et al. [80], Voronoi tessellation of receptors to reveal clefts by Lewis [81], solid modeling techniques to display cavity-like binding regions in proteins by Ho and Marshall [82], and the various measures of molecular shape put forward by Hopfinger and Burke [83].

6 Equivalence Relationships

As we have already intimated, chemistry is a science that makes very considerable use of categories or classes of objects. The classification of the objects and phenomena of the natural world merely formalizes what humans have been doing since time immemorial: trying to make sense of an overwhelmingly complex external environment by putting things into classes. Our realization that the classes we generate will likely be fuzzy and unstable with respect to time in no way diminishes our powerful impulse to systematize and thereby make manageable our experience of the world. Interestingly, several measures of classification performance have recently been proposed [84] to give us some idea of how well we operate. In chemistry, and in most other disciplines, the number of possible classifications we might generate is virtually infinite. Because most of these classifications would not be meaningful, it is necessary to restrict our field of discourse. We shall focus on specifically those classes that are defined in terms of one attribute only and that are quasi-stable in time. Even then, we shall be confronted with certain difficulties. In attempting to define the class of all inorganic compounds, for instance, problems arise as to the inclusion of borderline cases such as certain organometallics or cluster compounds. Any attempt to classify substances according to bonding type would lead us to additional difficulties, as will be evident from Fig. 6, which illustrates the gradual transitions that occur in bond types in various materials.

Classifications that enable us to partition chemical entities into mutually disjoint sets, that is to divide them up in such a way that each entity belongs to one and only one subset, are clearly of fundamental importance to chemists. In mathematical terminology this situation is described by saying that an equival-

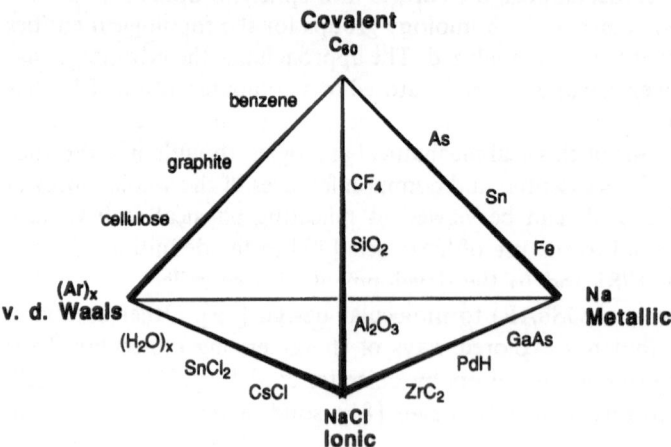

Fig. 6. An illustration of the gradual and almost continuous nature of the transitions that occur in the bond types of various chemical species

ence relation may be defined on the set in question. Such a statement implies that the members in any equivalence class formed by applying the relation will all be similar with respect to the defining relationship. An example of an equivalence relation is the assertion 'has the same numbers of protons as' which may be used to separate all known atoms into element classes. The members s_n (where $n = 1, 2, \ldots$) of a set S are said to satisfy an equivalence relation R_{eq} if they meet the following three requirements:

(i) Reflexivity, i.e. $s_x R_{eq} s_y$ holds for all $s_n \in S$;
(ii) Symmetricity, i.e. if $s_x R_{eq} s_y$, then $s_y R_{eq} s_x$; and
(iii) Transitivity, i.e. if $s_x R_{eq} s_y$ and $s_y R_{eq} s_z$, then $s_x R_{eq} s_z$.

The equivalence relation is both easy to comprehend and relatively straightforward to apply, though it must be cautioned that it is seldom rigorously applicable in practice. For this reason it has come to be regarded as a rough kind of standard against which less rigorously defined similarity relationships might be measured. The history of modern chemistry has frequently involved the search for new equivalence relationships. A listing of some of the more important ones discovered over the past 200 years is presented in Table 3.

Having now defined the equivalence relation, we are in a position to consider several other kinds of similarity that may be thought of as modified versions of the equivalence relation. By either weakening or strengthening the three defining requirements of reflexivity, symmetricity, and transitivity, it becomes possible to obtain a host of new relations, each of which characterizes a new variety of similarity. To keep our discussion within manageable proportions, however, we shall describe only one example each of the result of weakening and of the strengthening of these requirements. The weakening of the equivalence relation requirements leads to so-called orderings, that is to say they impose some kind of order on set members short of assigning them to disjoint classes. Suppose we weaken the requirement of symmetricity. This requirement can be weakened in three distinct ways, namely by the imposition of:

(i) Asymmetry, which implies that no set members $s_n \in S$ satisfy the relationships $s_x R s_y$ and $s_y R s_x$;
(ii) Antisymmetry, which implies that, if the relationships $s_x R s_y$ and $s_y R s_x$ are satisfied, then $x = y$; and
(iii) Connectedness, which implies that pairs of set members s_x, $s_y \in S$ satisfy either the relationship $s_x R s_y$ or $s_y R s_x$ if $x \neq y$.

In cases (i) and (ii) the relationship R is said to generate a partially ordered set or poset, whereas in case (iii) a totally ordered set is produced. Let us now move on to some chemically relevant instances.

Such weaker relationships are usually represented in the physical sciences by so-called Hasse or chain diagrams. These are constructed by depicting set members, s_n, as points, a_n, in some geometrical space and then connecting pairs of points that satisfy the relationship R. Chain diagrams have been widely employed in the chemical context to order members of chemical series, such as

Table 3. A listing of the 18 isomers of the molecule of octane with the number of graph paths of lengths 1 through 7 for each of the structures

Name of molecule	Graph	Number of paths p_i of length i						
		P_1	P_2	P_3	P_4	P_5	P_6	P_7
2,2,3,3-Tetramethylbutane		7	12	9				
2,2,4-Trimethylpentane		7	10	5	6			
2,2,3-Trimethylpentane		7	10	8	3			
2,3,3-Trimethylpentane		7	10	9	2			
2,3,4-Trimethylpentane		7	9	8	4			
2,2-Dimethylhexane		7	9	5	4	3		
3,3-Dimethylhexane		7	9	7	4	1		
2,5-Dimethylhexane		7	8	5	4	4		
2,4-Dimethylhexane		7	8	6	5	2		
2,3-Dimethylhexane		7	8	7	4	2		
3-Methyl-3-Ethylpentane		7	9	9	3			
2-Methyl-3-Ethylpentane		7	8	8	5			
3,4-Dimethylhexane		7	8	8	4	1		
2-Methylheptane		7	7	5	4	3	2	
3-Methylheptane		7	7	6	4	3	1	
4-Methylheptane		7	7	6	5	2	1	
3-Ethylhexane		7	7	7	5	2		
n-Octane		7	6	5	4	3	2	1

homologous or congeneric series. Orderings made on the basis of the structural characteristics of the molecular graphs of the species concerned often reflect remarkably well correlations in the physical and chemical behavior of the species. Thus, Randić and Wilkins [84] demonstrated that the number of paths of different lengths in the alkane isomers could be used to order these isomers according to a range of properties, such as their critical volume or heat of combustion. A typical partial ordering for the 18 isomers of octane in terms of paths of length two and three is shown in Fig. 7. Note that the connected points

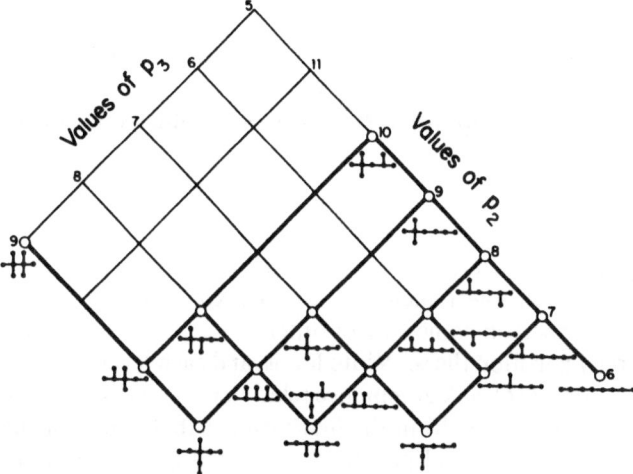

Fig. 7. A Hasse diagram depicting the 18 isomers of octane with a partial ordering based on the number of paths of length two (p_2) and of length three (p_3) for each isomer. The diagram derives from the work of Randic and Wilkins [84]

from a topological graph in which the points are in a sequence: a point a_1 that precedes point a_2 is always drawn below or to the left of point a_2. Whenever pairs of points, say (a_x, a_y) are not comparable because $x = y$, the points are drawn at the same height and may be connected with an equals sign. In Fig. 7 only the pairs of path lengths (p_2, p_3) are taken into account instead of the septuples of path lengths that might have been considered in a more exhaustive treatment (see Table 3). As a result of this restriction to pairs, two of the isomeric structures are not comparable, i.e. they possess identical (p_2, p_3) values. Even so, this ordering still reflects the trend in critical densities and pressures, apart from one discordant pair which is attributed to experimental error. The same kind of outcome was found when this approach was extended to the nonanes [85] and the decanes [86].

Just as partial ordering results from a weakening of the equivalence relation on a set, hierarchical ordering is achieved by a strengthening of the relation through the imposition of additional structure. Because of its powerful structuring potential, hierarchical ordering is one of the most significant of all the methods of classification and is widespread in its applications. Virtually all complex systems are hierarchically ordered and living systems afford prime examples. In fact, in living systems, the life process is now viewed as residing not in the so-called living matter itself but rather in the hierarchical organization and interaction of this matter [87]. Mathematically, a hierarchical structure is a set with a superimposed antisymmetric and dominance relationship. Thus, the hierarchy H imposed on the set S may be represented as the ordered triple $H = \langle S, s_n, D \rangle$, where $s_n \in S$ and D form the dominance relation. In addition, the following three criteria must be satisfied:

(i) In the hierarchical ranking of the s_n, the process always starts from the first member, s_1:
(ii) The member s_1 is always in relationship D to each of the other s_n members; and
(iii) For all $s_n \in S$ except s_1, precisely one member of S will satisfy the relationship $s_1 D s_n$ for all $n > 1$.

The relation D gives rise to either control hierarchies, in which restraint is exerted by higher members on lower members, or to taxonomic hierarchies which simply rank the members in some sort of order.

Both types of hierarchy partition the set S into levels, and each level typically comprises a collection of subsets of S. A dominance relation thus has the effect of ranking every member of a set, apart from the first member that starts off the hierarchy. A hierarchically nested set of partitions generated by the action of some dominance relation can be represented as a graph-theoretical tree. Such a tree normally has as its root the set member s_1. Trees may be labeled or unlabeled as required and no vertex is of degree greater than three, with the possible exception of the root. A level L_x in the tree is said to rank higher than some other level L_y if and only if $x < y$. If a specific height is associated with each internal vertex, a tree is said to be a *dendogram*. The 20 naturally occurring amino acids can be represented in the form of a dendogram in which the relative height of any

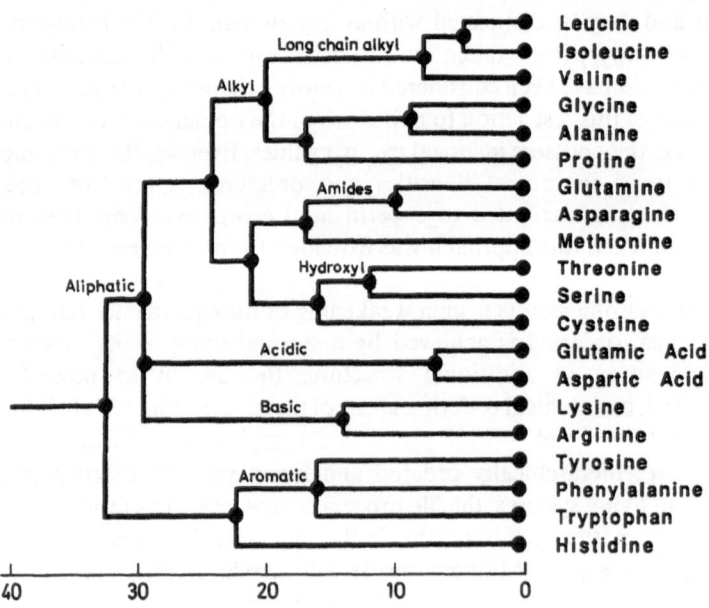

Fig. 8. A dendogram of the 20 naturally occurring amino acids displaying the hierarchical similarity in their chemical properties. The figure derives from the work of Sneath [88]

vertex affords a measure of the similarity of the amino acid represented. Based on a method developed by Sneath [88], the amino acids may be classified hierarchically as shown in Fig. 8. The relationships revealed by this device accord with our intuitive assessment of the behavior of these acids, though this fact has not stopped various workers from trying to improve on this means of classification. In particular, attempts have been made to render the hierarchical organization of biological systems quantitative by the imposition of conditions necessary for the emergence of functional hierarchies [89], and by exploring the possibility of revealing hierarchical structure with the aid of neural networks [90].

7 Scaling and Self-Similarity

Although the term scaling has come to be associated with a variety of different meanings, all of them pertain in some way to the differing kinds of invariance that are exhibited in the behavior of physical or biological systems. Invariance is encountered whenever a specific set of transformations carried out on some system leave certain features of that system unchanged. The current emphasis on what is left unaltered in a system after the performance of one or more operations on the system is an entirely modern one, since the existence of invariance has been known for several centuries. For example, in the chemical domain, early instances of invariance are to be found in the Law of Constant Proportions dating from 1804 [91] and the Law of Multiple Proportions dating from 1810 [92]. The former law states that the masses of the constituents of homogeneous substances bear a constant ratio to each other and to the mass of the compound, whereas the latter law sates that if two substances combine in more than one ratio, then the masses of the first substance that combine with a fixed mass of the second substance will be in a simple ratio. A listing of some of the major kinds of invariance that are associated with various chemical entities is presented in Table 4. This table reveals that the exploitation of invariances has resulted in the discovery of both general laws and the development of numerous significant practical applications. Because the presence of an invariance in a system is now interpreted as effectively defining a law for that system [93], scaling relationships have tended to assume an ever increasing prominence in recent years.

In fact, so important have invariances become that Rosen [94] has argued that they are essential for the performance of science. His reasoning was based on an analysis of the analogy (which we have already discussed in Sect. 4). Rosen [94] views the analogy as an invariance of some relation or statement that continues to hold when elements involved in that analogy are changed, a situation he describes as the symmetry of the analogy. This particular kind of symmetry lies at the very foundations of science, primarily because both reproducibility and predictability are interpreted as symmetries of analogy. In

Table 4. A listing of some of the major kinds of invariance that are associated with a variety of chemical entities

Chemical system	Invariance	Relevant parameter	Application
Molecule	Rotation/reflection	Molecular point group	Spectroscopy
Crystal lattice	Spatial translation	Space group	X-ray analysis
Homogeneous surface	Surface translation	2-dimensional unit cell	Adsorption studies
Lattice defect	Homotopy	Burgers vector	Crystal properties
Irregular structure	Dilation (self-similarity)	Fractal dimension	Scaling laws

the case of reproducibility, any experimental input into a system will generate a computable output, whereas, in the case of predictability, any experimental input can be substituted for any other input within the same system to yield a known result. This is possible only because both of these concepts are based on the existence of an invariant analogical relationship between the experimental input and the resulting output. Such a relationship is usually expressed in mathematical form and in many instances is referred to as a scientific law. Thus, in a very real sense, scientific laws are simply statements of similarity. Although laws are formally regarded as statements of universal validity that characterize some aspect of objective reality, they actually tell us that all the members of some equivalence class exhibit analogous (or possibly identical) behavior under prescribed conditions. It is thus hardly surprising that Jensen [95] concluded that "it cannot be denied that analogy plays an important role in scientific creativity."

Over the past two decades numerous scaling relationships have been established for a whole range of entities of importance to chemists, including molecular species of various kinds, reaction surfaces, powders, three-dimensional solid structures, and even flames. Interest in this area first began to blossom after the appearance in 1977 of the now classic book *Fractals: Form, Chance and Dimension* by Mandelbrot [96]. This work pointed out that virtually all of the objects in the real world are fractals, that is to say they are irregular, rugged, fragmented or tangled. In spite of their apparent imperfection, however, such objects display a hidden invariance that is usually described as self-similarity [97]. This type of invariance manifests itself when fractal objects become magnified in size, for it is found that upon dilation the overall geometric features of the objects remain unchanged. Fractal objects accordingly possess scale-invariant structures, and such invariance typically holds for progressive magnifications of the structures until roughly atomic-sized features become visible to the naked eye. The most common way of characterizing self-similar structures is in terms of their fractal dimensionaltiy. This parameter, which we represent by the symbol d_f, is the exponent that relates the mass of a fractal object, M, to its size, Q, via the relationship

$$M \sim Q^{d_f}.$$

Fractal dimensionalities in most cases assume values that lie in the range $1 \leq d_f \leq 3$.

One caveat that should be mentioned at this point is that fractal dimensionalities are not to be understood as geometrical or Euclidean dimensions. They are to be interpreted rather as a measure of the kinkiness or tortuousness of physical objects. It is for this reason that a highly twisted and bent one-dimensional object such as a wire can have a fractal dimensionality that exceeds two [98]. In the chemical context, fractal concepts have been exploited in at least three major areas of application. First, studies have been carried out on, and determinations have been made of, the fractal dimensionalities of many large molecules, such as protein molecules [99] and long-chain alkane species [100]. Second, a wide variety of surfaces have been investigated, including those with complex morphologies [101], the exposed outer surfaces of dendritic polymers [102], the solvent-accessible surfaces of biomacromolecules [103], and chemically reactive surfaces [104, 105]. Third, a somewhat miscellaneous group of three-dimensional objects have been studied, and these have included fractal powders and clusters [106], fractal turbulent flames [107], fractal reaction kinetics [108], and even fractal foodstuffs [109]. Since the surfaces of molecules and solid-state materials play such a crucial role in many of the phenomena and processes of chemistry, e.g. in adsorption processes in heterogeneous systems, in catalysis and biocatalysis, in many transport processes, and especially in the interactions now generally designated as molecular recognition, it is of vital importance that appropriate measures are becoming available to characterize these surfaces.

A novel method of deriving a measure of the similarity of two molecules that depends on scaling was put forward by Mezey [42]. The measure, referred to as a scaling-nesting similarity measure, is actually a scaling factor and is the amount of reduction in the size of a molecule that is necessary for it to fit entirely inside the volume of another molecule. The reduction process is illustrated in Fig. 9 for two two-dimensional molecules M and N. In the case that the two molecules are of identical volume, the scaling factor immediately affords a measure of the similarity of the species: the larger the scaling factor the greater the similarity will be. In the more usual case where the initial molecular volumes differ, it is

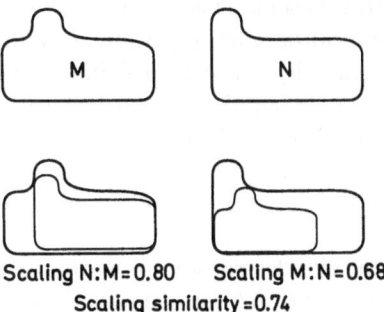

Scaling N:M=0.80   Scaling M:N=0.68
Scaling similarity =0.74

Fig. 9. A depiction of the scaling that is necessary to get a pair of two-dimensional molecules to fit exactly into the area taken up by one of the pair. This diagram derives from the work of Mezey [42]

necessary to scale one of the molecules down until it fits precisely into the volume occupied by the other molecule. In general, to achieve an optimal fit, the molecule being scaled down can be repositioned or rotated as required. Differing similarity measures can be defined according to which kinds of motion are permitted. For instance, if rotations are not allowed, the similarity measure is referred to as a constrained shape similarity measure, whereas if the centers of mass of the two molecules are fixed but rotations are allowed a position-dependent similarity measure is obtained. Since a different scaling may be required to fit M into N from that required to fit N into M, the two measures that are obtained are designated semi-similarity measures; the overall similarity is determined by simply taking the arithmetic mean of the two measures in such cases. These similarity measures are likely to prove of value in the study of chiral molecules as the measures can be directly related to the perception of the prominence of chirality in molecular species [42]. Although they are too new to have been used yet in any significant application, they would seem to hold considerable promise for the future.

8 Similarity of the Future

In this introduction to similarity we have emphasized the various uses of similarity for the characterization and comparison of molecular structures. Up to this point our focus has been primarily on the past and the present. We have discussed both early and contemporary concepts of similarity, we have outlined the various major kinds of similarity that have been exploited in the chemical context, and we have examined a wide range of differing measures of molecular similarity. The time has now come to attempt a peek into the future. This is, of course, a hazardous venture since, but its very nature, the future is largely unpredictable. However, some trends have begun to emerge in recent years and it is to these that we turn to guide us on our journey into the unknown. Of one thing we can be virtually certain: many new measures of molecular similarity will continue to be developed as the concept of similarity applied at the molecular level is likely to be of enduring importance. We have already witnessed its major importance in the area of pharmaceutical drug design. This importance will probably grow substantially over the next several decades, especially as similarity measures and similarity approaches generally are extended into ever newer areas. These areas may well encompass the design of new agrochemicals, such as environmentally friendly pesticides and herbicides, fuels that have close to optimal efficiency under prescribed operating conditions, and replacement body fluids – such as the vitreous humor in the eye or artificial blood – that are really effective.

The sophistication of the approaches to the description of molecular similarity will also likely increase notably in the future. Within the space of the past two decades, for instance, we have seen the description of molecular shape progress from relatively primitive methods that relied on the silhouettes of molecules viewed from different angles to shape group methods, dynamic shape analysis, the use of shape codes, shape matrices, and shape globe invariance maps [42]. The use of quantum-chemical methods has also grown by leaps and bounds. Starting from the first quantum-chemical measure of molecular similarity put forward by Carbó et al. [68] in 1980, we have now reached the point where there are numerous such similarity measures based on electron densities, electrostatic potentials, and molecular electric fields [110]. In an analogous fashion our understanding of the molecular recognition process has evolved from the embryonic concept of a lock and key into models that allow for such things as dynamic shape changes at the molecular level, the matching of charge distributions, and hierarchical involvement of the various structural features involved in the docking process [57]. In seems probable that this surge ahead in recent years will be paralleled by even more dramatic developments well into the next century. This prognostication would seem to be confirmed by the exponential rise in the number of publications on the subject – ranging from books to scientific papers – that have appeared during the past decade.

Throughout our discussion of similarity we have endeavored to make it clear that the use of similarity confers mixed blessings because the concept has the nature of a double-edged sword. On the one hand, similarity is an extremely valuable and exciting new tool in the chemist's armamentarium, whereas on the other hand the concept cannot be defined in absolute terms and thus is often difficult to express in explicit terms. For instance, the classes that are formed by grouping together molecules or other chemical entities in terms of their similarity are not in general describable by crisp sets. They will almost always be soft-edged and so have a certain fuzziness associated with them. From our earlier mention of this theme, we know that both similarity-based concepts and classes can be characterized in terms of fuzzy logic [18]. Although this state of affairs may at first sight seem restricting and undesirable, it has been increasingly exploited in recent years and turned to considerable advantage – a process that will likely continue and even accelerate in the years ahead. Fuzzy concepts and classes can be processed by the use of neural networks [111, 112]. The latter are software devices that are being increasingly used to obtain definitive answers from data that are incomplete, ill-defined, or fuzzy. Given this remarkable capability, it is hardly surprising that neural networks are rapidly gaining in acceptance and acclaim [111]. Learning to live with fuzzy concepts and fuzzy sets therefore need not be interpreted as accommodation to an unwelcome and anomalous situation. It should be seen rather as an exciting and growing opportunity that affords a radically new approach to data handling. That neural networks will play a significant role in the future of similarity studies cannot be doubted. In fact, we believe it is only a matter of time until "almost everybody knows that, outside of mathematics and logic, all definitions are fuzzy," to quote the words of Martin Gardner [113]. When this new awareness comes about, the time will be ripe for a massive exploitation of neural networks. They will doubtless be used as a springboard that will enable us to make major studies forward and also see us well into the twenty-first century.

9 References

1. Rouvray DH (1990) In: Johnson MA, Maggiora GM (eds) Concepts and applications of molecular similarity. Wiley-Interscience, New York, chap 2, p 15
2. The idea that science is based on a novel way of thinking that runs counter to common sense is brought out very clearly in the book of Cromer A (1993) Uncommon sense. Oxford University Press, New York Oxford, esp chap 1, p 3
3. Rouvray DH (1992) J Chem Inf Comput Sci 32: 580
4. Warrington J (ed) (1965) Plato's Timaeus. Dent, London, Sect 31C, p 22
5. Rouvray DH (1994) J Chem Inf Comput Sci 34: 446
6. Rouvray DH (1992) J Mol Struct (Theochem) 259: 1
7. Gotthelf A (ed) (1985), Aristotle on nature and living things. Mathesis, Pittsburgh, esp p 95
8. McKeon R (ed) (1941), The basic works of Aristotle. Random House, New York, p 17
9. See, for instance the collection of works in Harnad, S. (ed) (1987), Categorical perception: The groundwork of cognition. Cambridge University Press, Cambridge New York, esp chap 16, p 455
10. Medin DL (1989) Am Psychol 44: 1469
11. Oden GC, Lopes L (1982) In: Yager RR (ed) Recent developments in fuzzy set and possibility theory. Pergamon, Elmsford, New York, p 75
12. DaCosta NCA, French S (1990) Phil Sci 57: 248
13. Bateson G (1979) Mind and nature: A necessary unity. Dutton, New York, p 38
14. Lakoff G (1987) Women, fire, and dangerous things: What categories reveal about the mind. University of Chicago Press, Chicago London, esp chap 1, p 5
15. Thagard P (1990) Synthese 82: 255
16. Medin DL, Barsalou LW (1987) In: Harnad (ed) Categorical perception. Cambridge University Press, Cambridge, UK, p 455
17. Jones A, Kaufmann A, Zimmermann H-J (1986) Fuzzy sets theory and applications. Reidel, Dordrecht Boston
18. Yager RR, Zadeh LA (eds) (1992), An introduction to fuzzy logic applications in intelligent systems. Kluwer, Boston Dordrecht London, esp chap 1, p 1
19. Zadeh LA (1965) Inform and control 8: 338
20. Tversky A (1977) Psychol Rev 84: 327
21. Krueger LE (1978) Psychol Rev 85: 278
22. Freeman WJ (1991) Sci Am 264: 78 [Feb]
23. Livingston M, Hubel D (1988) Science 240: 740
24. Smith LB (1989) In: Vosniadou S, Ortony A (eds) Similarity and analogical reasoning. Cambridge University Press, Cambridge. UK, chap 5, p 146
25. Sensi P (1982) In: Bindra JS, Lednicer D (eds) Chronicles of drug discovery. Wiley-Interscience, New York, chap 9, p 201
26. Gerhardt CF (1845) Ann Chim 14: 117
27. Laurent A (1854), Méthode de chimie. Mallet-Bachelier, Paris, p 373
28. de Maupertuis PLM (1744) Mém. Acad Roy Sci, Paris, 546
29. Le Chatelier H (1888) Ann Mines 13: 157
30. Döbereiner JW (1829) Ann Phys Chem 15: 301
31. van Spronsen JW (1969), The periodic system of the chemical elements. Elsevier, Amsterdam
32. Rouvray DH (1944) Chem Brit 30: 373
33. Kong F (1982) J Mol Struct 90: 17
34. Hefferlin R (1989) Periodic systems of molecules and their relation of the systematic analysis of molecular data. Mellin Press, Lewiston, New York, esp chap 12, p 414
35. Ibid, idem, chap 11, p 396
36. Dias JR (1993) J Chem Inf Comput Sci 33: 117

37. Dias JR (1993) Tetrahedron 49: 9207
38. Rouvray DH (1993) New Sci 138: 35
39. Rouvray DH (1987) J Comput Chem 8: 470
40. Rouvray DH (1986) Sci Am 245: 40
41. Kier LB, Hall LH (1986) Molecular connectivity in structure-activity analysis. Research Studies Press, Chichester, UK
42. Mezey PG (1993) Shape in chemistry. VCH Press, New York Weinheim Cambridge, esp chap 4, p 82
43. Bertz SH (1981) J Am Chem Soc 103: 3599
44. Randić M (1991) J Comput Chem 12: 970
45. Johnson MA (1989) J Math Chem 3: 117
46. Willett P (1987) Similarity and clustering in chemical information systems. Research Studies Press, Letchworth, UK, chap 3, p 89
47. Willett P, Winterman V (1986) Quant Struct-Act Relat 5: 18
48. Gould SJ (1987) An urchin in the storm. Norton, New York London, chap 16, p 221
49. VanLehn K (1991) Cognit Sci 15: 1
50. Clement CA, Gentner D (1991) Cognit Sci 15: 89
51. Primas H (1982) Chimia 36: 293
52. Greenberger DM (1982) In: Shimony A, Feshbach H (eds) Physics as natural philosophy. MIT Press, Cambridge MA London, p 178
53. Bohr N (1928) Nature (London) 121: 580
54. Murdoch D (1987) Niels Bohr's philosophy of physics. Cambridge University Press, Cambridge UK, chap 4, p 59
55. Fischer E (1894) Chem Ber 27: 2985
56. Pattee HH (1979) Biosystems 11: 217
57. Greenspan NS (1992) Bull Inst Pasteur 90: 267
58. Rebek J (1987) Science 235: 1478
59. Meng EC, Shoichet BK, Kuntz ID (1992) J Comput Chem 13: 505
60. DesJarlais RL, Sheridan RP, Seibel GL, Dixon JS, Kuntz ID, Venkataraghavan (1988) J Med Chem 31: 722
61. Willett P (1991) Three-dimensional chemical structure handling. Research Studies Press, Taunton, UK, esp chap 1, p 1
62. Kenakin TP (1989) Trends Pharmacol Sci 10: 18
63. Schneider H-J (1991) Angew Chem Int Ed Engl 30: 1417
64. Shoichet BK, Kuntz ID (1993) Protein Engineering 6: 723
65. Motoc I (1983) Topics Curr Chem 114: 94
66. Cano FH, Martínez-Ripoll M (1992) J Mol Struct (Theochem) 258: 139
67. Mezey PG (1987) Int J Quant Chem: Quant Biol Symp 14: 127
68. Carbó R, Leyda L, Arnau M (1980) Int J Quant Chem 17: 1185
69. Hodgkin EE, Richards WG (1987) Int J Quant Chem: Quant Biol Symp 14: 106
70. Ponec R (1984) Coll Czech Chem Commun 49: 455
71. Ponec R (1987) Coll Czech Chem Commun 52: 555
72. Ponec R, Strnad M (1992) Int J Quant Chem 42: 501
73. Cooper DL, Allan NL (1989) J Comput-Aided Mol Design 3: 253
74. Allan NL, Cooper DL (1992) J Chem Inf Comput Sci 32: 587
75. Woolley RG (1991) J Mol Struct (Theochem) 230: 17
76. Amann A (1992) S Afr J Chem 45: 29
77. Arteca GA, Mezey PG (1989) J Math Chem 3: 43
78. Mezey PG (1993) J Math Chem 12: 365
79. Mezey PG (1992) J Chem Inf Comput Sci 32: 650
80. Leicester SE, Finney JL, Bywater RP (1988) J Mol Graphics 6: 104
81. Lewis RA (1989) J Comput-Aided Mol Design 3: 133
82. Ho CMW, Marshall GR (1990) J Comput-Aided Mol Design 4: 337
83. Hopfinger AJ, Burke BJ (1990) In: Johnson MA, Maggiora GM (eds) Concepts and applications of molecular similarity. Wiley-Interscience, New York, chap 7, p 173
84. Randić M, Wilkins CL (1979) J Phys Chem 83: 1525
85. Randić M, Wilkins CL (1979) Chem Phys Lett 63: 332
86. Randić M, Trinajstić N (1982) Math Chem 13: 271
87. Zylstra U (1992) Synthese 91: 111

 88. Sneath PHA (1966) J Theor Biol 12: 157
 89. Totafurno J, Lumsden CJ, Trainor LEH (1980) J Theor Biol 85: 171
 90. Sakaguchi H (1989) Prog Theor Phys 82: 321
 91. Proust JL (1804) J Phys 59: 321
 92. Berzelius JJ (1810) Ann Phys 35: 269
 93. Barrow JD (1988) The world within the world. Clarendon Press, Oxford, chap 3, p 86
 94. Rosen J (1990) Symmetry 1: 19
 95. Jensen WB (1986) Comput Math Appl 12B: 487
 96. Mandelbrot BB (1977) Fractals: Form chance and dimension. Freeman, San Francisco. This
     work has been followed by an avalanche of other books dealing with the concept of self-
     similarity applied in the chemical context.
 97. An excellent introduction to this area is to be found in the book by Kaye BH (1989) A random
     walk through fractal dimensions. VCH, Weinheim, Germany
 98. Musès C (1991) In: Rassias GM (ed) The mathematical heritage of CF Gauss, World Scientific
     Publ Co. Singapore, p 526
 99. Wagner GC, Colvin JT, Allen JP, Stapleton HJ (1985) J Am Chem Soc 107: 5589
100. Rouvray DH, Pandey RB (1986) J Chem Phys 85: 2286
101. Avnir D, Farin D, Pfiefer P (1992) New J Chem 16: 439
102. Farin D, Avnir D (1991) Angew Chem Int Ed Engl 30: 1379
103. Zachmann C-D, Kast SM, Sariban A, Brickmann J (1993) J Comput Chem 14: 1290
104. Pfeifer P (1985) Chimia 39: 120
105. Farin D, Avnir D (1987) J Phys Chem 91: 5517
106. Chapters 5, 7 and 9 of the book by Kaye cited in ref. 97 contain much relevant information on
     this theme
107. Gouldin FC, Bray KNC, Chen J-Y (1989) Combust Flame 77: 241
     108. Kopelman R (1988) Science 241: 1620
109. Peleg M (1993) Crit Rev Food Sci Nutr 33: 149
110. Hodgkin EE, Richards WG (1987) Int J Quant Chem: Quant Biol Symp 14: 105
111. Rouvray DH (1993) Chem Brit 29: 495
112. Zupan J, Gasteiger J (1993) Neural networks for chemists: An introduction. VCH Press,
     Weinheim, Germany
113. This is a comment made by Martin Gardner in a book review that appears on the dust jacket of
     McNeill D, Freiberger P (1993) Fuzzy logic. Simon and Schuster, New York

Foundations and Recent Developments on Molecular Quantum Similarity*

Emili Besalú, Ramon Carbó**, Jordi Mestres and Miquel Solà

Institut de Química Computacional, Universitat de Girona, Albereda, 3–5, 17071 Girona (Spain)

Table of Contents

---

* A contribution of the Grup de Química Quàntica de l'Institut d'Estudis Catalans.
** To whom correspondence should be addressed.

A general definition of the Quantum Molecular Similarity Measure is reported. Particular cases of this definition are discussed, drawing special attention to the new definition of Gravitational-like Quantum Molecular Similarity Measures. Applications to the study of fluoromethanes and chloromethanes, the Carbonic Anhydrase enzyme, and the Hammond postulate are presented. Our calculations fully support the use of Quantum Molecular Similarity Measures as an efficient molecular engineering tool in order to predict physical properties, biological and pharmacological activities, as well as to interpret complex chemical problems.

1 Introduction

In our laboratory a progressive development has been made in successive papers [1, 2] during the past twelve years in order to construct a rigorous definition of Molecular Quantum Similarity Measures (MQSM) and establish their quantum mechanical meaning. The present theoretical scheme constitutes an exposition of the most important MQSM we are using in our laboratory. This kind of Measures are implemented in the QMOLSIM [3] program system, where the most recent ideas developed on this topic and the relevant part of the formulation quoted in [2] are contained. Here, special emphasis has been made with respect to the so called Gravitational Integrals. Their definition and practical computation in a MO LCAO framework are described.

2 Density Integral Transforms (DIT) and Molecular Quantum Similarity Measures (MQSM)

Although density matrix elements are the natural candidates to be used in MQSM, a more general framework can be described. The concept of DIT cannot be ignored for this purpose and this section is devoted to the establishment of a theoretical foundation.

2.1 Definition

Let us define an n-th order Density Integral Transform as the transformation of a density matrix element [4], $\rho^{(n)}(r, u, p)$, of the same order:

$$P^{(n)}(r, s, p) = \int \Omega(r, s, u, p) \rho^{(n)}(r, u, p)\, du, \tag{1}$$

where the operator $\Omega(r, s, u, p)$ is the kernel of the transform. r, s and u are electronic coordinates and the optional vector p provides the dependence of the density matrix element, the operator or both with respect to a parameter set.

2.2 Particular Cases of DIT

When the operator Ω appearing in Eq. (1) is defined as $\Omega(r, s, u, p) = \delta(u - s)$, then the DIT becomes $P^{(n)}(r, s, p) = \rho^{(n)}(r, s, p)$ and the transformation leaves invariant the density matrix element. If Ω is defined as $\Omega(r, s, u, p) = \delta(u - r)$, then $P^{(n)}(r, s, p) = \rho^{(n)}(r, p)$ and the transformation returns a diagonal element of the density matrix: the n-th order density function. When within the DIT definition the operator form $\Omega(r, s, u, p) = |u - s|^{-1}$ is used, then $P^{(n)}(r, s, p) = V^{(n)}(r, s, p)$. This is: a generalized form of the electrostatic potential is obtained. One can call this general formulation an n-th order Electrostatic Potential. A particular case occurs when using the density function $\rho^{(1)}(s, p)$ in Eq. (1), when the resulting transform coincides with the usual electrostatic potential function.

2.3 Structure of MQSM

MQSM are integrals involving two or more DITs attached to molecular systems and an optional operator. In this part we will try to describe a general framework from which the similarity between quantum objects can be computed.

With all this considerations in mind, some MQSM integrals can be defined, in a very general framework [2] or in the more concrete practical way as follows.

The n-th order Quantum Similarity Measure between two quantum systems A and B with respect to an operator W, can be defined as an integral of the following kind:

$$Z_{AB}^{(n)}(W,p) = \iiiint W(r_1,r_2,s_1,s_2,p)$$
$$\times P_A^{(n)}(r_1,s_1,p)\,P_B^{(n)}(r_2,s_2,p)\,dr_1\,dr_2\,ds_1\,ds_2, \qquad (2)$$

where $P_A^{(n)}(r,s,p)$ and $P_B^{(n)}(r,s,p)$ are the DIT related to the molecules A and B respectively. W is an operator which is, usually, definite positive.

Particular cases of previous Eq. (2) are directly implemented in the QMOL-SIM program.

a) When W is defined as the following Dirac delta functions product:

$$W(r_1,r_2,s_1,s_2,p) = \delta(r_1 - r_2)\,\delta(s_1 - s_2) \qquad (3)$$

then, one is dealing with an Overlap-like MQSM:

$$Z_{AB}^{(n)}(\delta(r_1 - r_2)\,\delta(s_1 - s_2),p) = \iint P_A^{(n)}(r,s,p)\,P_B^{(n)}(r,s,p)\,dr\,ds. \qquad (4)$$

b) If W is a product of two Coulomb operators,

$$W(r_1,r_2,s_1,s_2,p) = [|r_1 - r_2||s_1 - s_2|]^{-1}, \qquad (5)$$

one obtains the following MQSM:

$$Z_{AB}^{(n)}(r_{12}^{-1} s_{12}^{-1},p) = \iiiint P_A^{(n)}(r_1,s_1,p)\,P_B^{(n)}(r_2,s_2,p)$$
$$\times [|r_1 - r_2||s_1 - s_2|]^{-1}\,dr_1\,dr_2\,ds_1\,ds_2, \qquad (6)$$

which can be considered an Electrostatic Potential MQSM.

c) When the involved DIT are density functions, the counterparts of Eqs. (4) and (6) read [1h, i]:

$$Z_{AB}^{(n)}(\delta(r_1 - r_2),p) = \int \rho_A^{(n)}(r,p)\,\rho_B^{(n)}(r,p)\,dr \qquad (7)$$

and

$$Z_{AB}^{(n)}(r_{12}^{-1},p) = \iint \rho_A^{(n)}(r_1,p)\,\rho_B^{(n)}(r_2,p)\,|r_1 - r_2|^{-1}\,dr_1\,dr_2 \qquad (8)$$

respectively. Equation (8) may be referred to as a Coulomb-like MQSM.

d) As another example, if the density matrix elements are transformed to n-th order electrostatic potentials and one uses a similar equation as Eq. (2) with the operator $W(r_1,r_2,r_3)$ definition:

$$W(r_1,r_2,r_3) = \tfrac{1}{4}\delta(r_3 - (r_1 + r_2)/2), \qquad (9)$$

then a similarity measure is obtained bearing *gravitational* operator structure:

$$Z_{AB}^{(n)}(\delta(r_3 - (r_1 + r_2)/2)/4,p) =$$
$$\iint \rho_A^{(n)}(r_1,p)\,\rho_B^{(n)}(r_2,p)\,|r_1 - r_2|^{-2}\,dr_1\,dr_2. \qquad (10)$$

This measure may be referred to as Gravitational-like MQSM.

e) One can consider any other system's appropriate DIT as the postive definite operator W in Eq. (2). A Triple Density Transform Similarity Measure (TDTSM) [1k] is then constructed. In terms of n-th order density matrix elements it reads, for example:

$$T_{AB;C}^{(n)} = \iiint \rho_A^{(n)}(r_1, r_2, p) \rho_C^{(n)}(r_2, r_3, p) \rho_B^{(n)}(r_3, r_1, p) \, dr_1 \, dr_2 \, dr_3. \tag{11}$$

2.4 Practical Implementation: LCAO Expressions of MQSM

Usually the DIT resulting from Eq. (1) are first order density functions. So we obtain in a MO LCAO framework:

$$\rho^{(1)}(r) = \sum_\mu \sum_\nu D_{\mu\nu} \chi_\mu(r) \chi_\nu(r), \tag{12}$$

where the parameter dependence in p has been omitted for simplicity's sake; $\{D_{\mu\nu}\}$ is the charge-bond order matrix and $\{\chi_\mu\}$ the AO basis set functions. When dealing with first order density matrix elements, Eq. (12) reads:

$$\rho^{(1)}(r, s) = \sum_\mu \sum_\nu D_{\mu\nu} \chi_\mu(r) \chi_\nu(s). \tag{13}$$

Using these definitions one can write, for example, the LCAO form of the measure of Eq. (2):

$$Z_{AB}^{(1)}(W) = \sum_{\mu, \nu} D_{\mu\nu}^A \sum_{\lambda, \sigma} D_{\lambda\sigma}^B$$
$$\times \iiint \chi_\mu^A(r_1) \chi_\nu^A(s_1) W(r_1, r_2, s_1, s_2) \chi_\lambda^B(r_2) \chi_\sigma^B(s_2) \, dr_1 \, dr_2 \, ds_1 \, ds_2. \tag{14}$$

In this way one can write the counterparts of previous defined MQSM in the MO LCAO framework.

a) An Overlap-like MQSM is written as:

$$Z_{AB} = \sum_{\mu, \nu} D_{\mu\nu}^A \sum_{\lambda, \sigma} D_{\lambda\sigma}^B \iint \chi_\mu^A(r) \chi_\nu^A(s) \chi_\lambda^A(r) \chi_\sigma^B(s) \, dr \, ds$$
$$= \sum_{\mu, \nu} D_{\mu\nu}^A \sum_{\lambda, \sigma} D_{\lambda\sigma}^B \langle A; \mu | \lambda; B \rangle \langle A; \nu | \sigma; B \rangle, \tag{15}$$

where the symbols $\langle A; \mu | \nu; B \rangle$ stand as overlap integrals between the atomic orbitals μ and ν of molecules A and B, respectively. It can be easily shown that the selfsimilarity integral Z_{AA} as Eq. (15), when applied in a closed shell case, returns $2N$ being N the number of electrons of molecule A. The NOEL method derived by Cioslowski and Fleischmann [5] for calculating MQSM is practically the same as Eq. (15).

b) An Electrostatic Potential MQSM can be similarly written as:

$$Z_{AB} = \sum_{\mu, \nu} D_{\mu\nu}^A \sum_{\lambda, \sigma} D_{\lambda\sigma}^B \langle A; \mu | r_{12}^{-1} | \lambda; B \rangle \langle A; \nu | r_{12}^{-1} | \sigma; B \rangle. \tag{16}$$

Here, the symbols $\langle A; \mu | r_{12}^{-1} | \lambda; B \rangle$ represent some sort of Coulomb-type integral involving only one atomic orbital of each molecule. The atomic orbitals μ and λ belong to molecules A and B respectively, then the integral is defined as:

$$\langle A; \mu | r_{12}^{-1} | \lambda; B \rangle = \iint \chi_\mu^A(r_1) |r_1 - r_2|^{-1} \chi_\lambda^B(r_2) \, dr_1 \, dr_2. \tag{17}$$

Finally, an Overlap-like MQSM involving first-order density functions can be written as:

$$\begin{aligned}
Z_{AB} &= \sum_{\mu, \nu} D_{\mu\nu}^A \sum_{\lambda, \sigma} D_{\lambda\sigma}^B \iint \chi_\mu^A(r) \chi_\nu^A(r) \chi_\lambda^B(r) \chi_\sigma^B(r) \, dr \\
&= \sum_{\mu, \nu} D_{\mu\nu}^A \sum_{\lambda, \sigma} D_{\lambda\sigma}^B \langle \mu\nu | \lambda\sigma \rangle,
\end{aligned} \tag{18}$$

where the terms $\langle \mu\nu | \lambda\sigma \rangle$ are four-center overlap integrals.

A Coulomb-like MQSM can be expressed in turn as:

$$Z_{AB} = \sum_{\mu, \nu} D_{\mu\nu}^A \sum_{\lambda, \sigma} D_{\lambda\sigma}^B \langle \mu\nu | r_{12}^{-1} | \lambda\sigma \rangle, \tag{19}$$

where use has been made of the well known repulsion integrals invovling orbitals of each molecule.

c) In a very similar manner, a Gravitational-like MQSM bears the following structure:

$$Z_{AB} = \sum_{\mu, \nu} D_{\mu\nu}^A \sum_{\lambda, \sigma} D_{\lambda\sigma}^B \langle \mu\nu | r_{12}^{-2} | \lambda\sigma \rangle. \tag{20}$$

Here, the integral definition holds:

$$\langle \mu\nu | r_{12}^{-2} | \lambda\sigma \rangle = \iint \chi_\mu^A(r_1) \chi_\lambda^B(r_1) r_{12}^{-2} \chi_\nu^A(r_2) \chi_\sigma^B(r_2) \, dr_1 \, dr_2. \tag{21}$$

In the Appendix, the evaluation of this kind of integrals is developed when the atomic orbitals are expanded in terms of GTO functions or linear combinations of them.

d) With respect to the TDTSM, as defined by means of the integral at Eq. (11), it can be computed as:

$$T_{AB;C} = \sum_{\mu, \nu} D_{\mu\nu}^A \sum_{\lambda, \sigma} D_{\lambda\sigma}^B \sum_{\alpha, \beta} D_{\alpha\beta}^C \langle A; \mu | \sigma; B \rangle \langle A; \nu | \alpha; C \rangle \langle B; \lambda | \beta; C \rangle, \tag{22}$$

where a similar notation as in Eq. (15) has been used. With respect to this Triple Density Transform, other possible forms can be constructed [1i].

Throughout this work, the optimization process of the relative position of the molecules when searching the maximal MQSM value between them has been done using Coulomb-like Similarities, see Eq. (19). This kind of MQSM has the advantage of reducing the core dependence, which is one of the main problems appearing when Overlap-like Similarities are computed. It can be shown [6] that, due to the presence of cusps of density in nuclei, once the mutual orientation of the two systems being compared has been optimized to maximize the MQSM, very small displacements of one system bring about radical changes in the value of the Overlap-like Similarity, making the optimization process fairly complicated. Fortunately, this problem practically disappears upon use of Eq. (19) to compute the MQSM.

3 Manipulation of MQSM and Visualization Techniques

All the ideas and definitions exposed here have been applied over Molecular Sets. As can be deduced from Eq. (2) of Sect. 2, all the MQSM can be collected into a matrix involving all the available pairs of molecules:

$$Z(W) = \{Z_{IJ}(W)\}, \tag{23}$$

where the superindex n has been omitted.

3.1 Similarity Indices

Once a set of quantum objects to study is chosen and the operator related to the MQSM definition in Eq. (2) is defined, the MQSM related to the set is unique. But they can be transformed or combined in order to obtain a new kind of auxiliary terms which can be named Quantum Similarity Indices (QSI). A vast quantity of possible MQSM manipulations leading to a variety of QSI definitions exists. The most used QSI are as follows.

a) The Cosine-like Similarity Index between two molecules I and J is constructed as:

$$C_{IJ}(W) = Z_{IJ}(W)[Z_{II}(W)Z_{JJ}(W)]^{-1/2}. \tag{24}$$

This cosine-like index has, for any pair of compared systems, a value between 0 (total dissimilarity) and 1 (complete similarity) depending on the similarity associated to the two molecules. Some authors refer to this index as Carbó Similarity Index.

b) The Distance Similarity Index, as shown in [1a], can also be redefined constructing a generalized Dissimilarity Index:

$$D_{IJ}(k, x, W) = [k(Z_{II}(W) + Z_{JJ}(W))/2 - xZ_{IJ}(W)]^{1/2}, \quad x \in [0, k]. \tag{25}$$

which for $k = x = 2$ reduces to a Euclidean distance index.

Also, the following index can be defined:

$$D_{IJ}(\infty, W) = \max(Z_{II}(W), Z_{JJ}(W)), \tag{26}$$

constituting the infinite order distance index.

c) The Hodgkin-Richards [7] or Tanimoto [8] indices can be cast into a generalized formula:

$$V_{IJ}(k, x, W) = (k - x)Z_{IJ}(W)D_{IJ}^{-2}(k, x, W), \quad k \in [0, 2], \tag{27}$$

where, when $k = 2$, with $x = 0$ the Hodgkin-Richards index is reproduced and when $x = 1$ the Tanimoto index is obtained.

d) The continuous form of the Petke index [9] can be defined as:

$$P_{IJ}(W) = Z_{IJ}(W) D_{IJ}^{-1}(\infty, W).$$ (28)

3.2 Extension of MQSM to Density Differences

In a recent work, Cioslowski et al. [10] defines the so called Fukui densities as the difference between the first order density function ρ^o of a system and either the first order density for the molecular positive or negative ion ρ^s ($s = (+)$, $(-)$). Let us define this density difference as:

$$\Delta^s = \rho^o - \rho^s.$$ (29)

Comparison of non-definite functions such as Δ^s, computed as a difference of two positive definite elements, can be done by means of the concept of a Signed Measure, applying the so called Hahn Decomposition Theorem [11].

Let us imagine two systems $\{A, B\}$ with known density function $\{\rho_I^o, \rho_I^s\}$ ($I = A, B$; $s = (+), (-)$). Then a Quantum Signed Similarity Measure can be defined between the Fukui density differences as:

$$
\begin{aligned}
F_{AB}^{ss'}(W) &= \int \Delta_A^s W \Delta_B^{s'} dr \\
&= \int (\rho_A^o W \rho_B^o + \rho_A^s W \rho_B^{s'}) dr - \int (\rho_A^s W \rho_B^o + \rho_A^o W \rho_B^{s'}) dr \\
&= (Z_{AB}^{oo}(W) + Z_{AB}^{ss'}(W)) - (Z_{AB}^{so}(W) + Z_{AB}^{os'}(W)).
\end{aligned}
$$ (30)

Optimal $F_{AB}^{ss'}(W)$ values can thus be obtained by means of maximal values of the four involved MQSM.

Signed measure similarity indices can be obtained in the usual way, by computing the signed selfsimilarity measures:

$$F_{II}^{ss}(W) = \int \Delta_I^s W \Delta_I^s dr = (Z_{II}^{oo}(W) + Z_{II}^{ss}(W)) - 2Z_{II}^{os}(W)$$ (31)

which is the square Euclidean distance between ρ_I^o and ρ_I^s.

In this sense, density differences can be compared in the same way as densities themselves.

3.3 Point Molecules and Molecular Point Clouds

Once a set of molecules is described by means of MQSM or QSI matrices, the problem arises on how to extract the information contained in the set. The way the authors apply these postulates to order the elements of the molecular set involves computing the Principal Components of the similarity matrix Z, which collect MQSM or QSI of the elements of the set. The eigenvectors of Z:

$$Z u_i = z_i u_i$$ (32)

form a set $U = \{u_i\}$ which can be used as a geometrical finite dimensional representation of the molecules. According to a well known variational principle, the eigenvectors of the definite positive matrix representing the set are the optimal space directions [12] where the projections of the objects can be made. This kind of projection ensures the minimal loss of information from the original set. Choosing representations in n eigenvectors, an n-dimensional space projection is obtained if it is necessary.

It is common practice to visualize the object set [1g,j] in order to achieve this goal. Using spatial multidimensional rotations, the final visualization can take care of the description of the set. An optimal visualization is searched in order to construct clusters allowing the whole set to be classified into subsets sharing a common range of a given property value. When dealing with a molecular set, the information which can be extracted from the object set is related to the structure-activity or structure-property relationships and may become particularly interesting are useful to solve molecular engineering problems.

3.4 Projection Techniques. Molecular Cloud Spread

The figures presented in this work are the projection of the resultant Point Clouds into bidimensional spaces. This enables us to grasp the information contained in the Molecular set in a finite dimensional human-like manner.

The matrix representation of a molecular set can be associated with a set of finite dimensional vectors representing the molecules. This point of view leads to the concept of Point-Molecules collected as a Molecular Point Cloud.

Several representations can be obtained since the matrix of the object coordinates may come from the eigenvectors of MQSM or QSI matrices of any kind.

Once a projection has been obtained a useful parameter can be defined: Molecular Cloud Spread, which measures the projected cloud ability to maintain their constituent points far away one from another [1e].

3.5 Mendeleev Postulates

From these previous considerations, a resumé can be structured in terms of four postulates, which we call Mendeleev Postulates in hommage to the first chemist who sought order between chemical substances.

The Mendeleev Postulates [2] state that it is always possible to extract information from the Molecular or Object set being studied. Briefly, they can be written as follows:

1) Every quantum system in a given state can be described by their DIT of the Density Matrix Elements.
2) Quantum systems can be compared by means of a MQSM or QSI.

3) Projection of a Quantum Object Set into some n-dimensional space is always feasible.
4) A Quantum Object Set Ordering exists.

Such conclusions lead to the Mendeleev Conjecture: "Object ordering induces order over the implicit relationships which exists between the set elements and their properties."

4 Approximate MQSM: Fitting Densities

4.1 Fitting Densities to Linear Combinations of Gaussian Functions

The field of Molecular Similarity has experienced remarkable progress in the last decade. The main topics covered by this area of research include Linear Free Energy Relationships (LFER) [13] and Quantitative Structure-Activity Relationships (QSAR) [14], although a large number of chemical definitions and concepts involve the Similarity Notion [15].

One of the most widely used definition of Molecular Similarity was originally introduced by one of us [1a,j] and it is defined in a quantum mechanical framework from the first order electron density functions $\{D_I, D_J\}$ of the two molecules being compared using some form of MQSM as described in Eq. (14).

The main drawback of most ab initio MQSM as defined in Sect. 2 is the need to calculate the four-centered integrals involved in Eq. (14) whose number scales to N^4. This problem becomes especially cumbersome when large molecules should be displaced one with respect to the other to maximize their Similarity which is equivalent to minimize the distance calculated by means of Eq. (25).

In order to circumvent this N^4 problem, a computational device has been designed [16]. In this method, the electronic first order density function is approximated by a linear combination of N Gaussian function set $\Gamma(r) = \{g_i(r)\}$:

$$\rho(r) \simeq D(r) = \sum_i^N c_i g_i(r). \tag{33}$$

Then, if we define the function difference as:

$$P(r) = \rho(r) - D(r) \tag{34}$$

one can optimally fit $D(r)$ to $\rho(r)$ provided that the quadratic error integral

$$\Delta(\Theta) = \iint P(r_1) W(r_1, r_2) P(r_2) \, dr_1 \, dr_2 \tag{35}$$

is minimized according to the least squares technique. If the coefficients of the expansion and the Gaussian function set $\Gamma(r)$ in Eq. (4) are collected into the column vectors c and the row vector $G(r) = (g_1(r), g_2(r), \ldots)$, respectively, that is:

$$D(r) = G(r)c \tag{36}$$

then c is given in a least squares sense by the matrix expression:

$$c = S^{-1}t, \tag{37}$$

where the elements of the vector t are defined by:

$$t_i = \iint \rho(r_1) W(r_1, r_2) g_i(r_2) \, dr_1 \, dr_2, \tag{38}$$

and the metric matrix S is constructed as:

$$S_{ij} = \iint g_i(r_1) W(r_1, r_2) g_j(r_2) \, dr_1 \, dr_2. \tag{39}$$

If $\Delta(\Theta)$ is minimized subject to the constraint that the equality:

$$\int D(r) \, dr = N_e \tag{40}$$

holds, then the expansion coefficients are given by the expression:

$$c = S^{-1}(t + \lambda n) \tag{41}$$

where

$$n_i = \int g_i(r) \, dr \tag{42}$$

and

$$\lambda = \frac{(N_e - n^T S^{-1} t)}{n^T S^{-1} n}, \tag{43}$$

N_e being the total number of electrons and λ a Lagrange multiplier, which is necessary to guarantee charge conservation.

Different fitting procedures can be conceived. For instance, the Overlap fitting equates $W(r_1, r_2)$ to $\delta(r_1 - r_2)$, whereas the Coulombic fitting takes $W(r_1, r_2)$ as r_{12}^{-1}. It is generally found [6] that Coulombic fitting is more expensive than Overlap fitting, without producing a much better description of the electron density from a MQSM point of view.

Within this fitting procedure, if N is taken as the number of Gaussian functions used in the expansion of Eq. (4) and n is the number of basis functions used to expand the wavefunction, evaluation of the Similarity using a fitted density results in an N^2-depending process, as opposed to an n^4-depending process when using the exact density.

4.2 MQSM with Fitted Densities

Previous work [16] has analyzed and discussed different fits to the exact electron density in terms of computational time used and ability to reproduce ab initio MQSM. In this study, MQSM have been carried out for a set of small molecules in order to show that MQSM, obtained from fitted densities, produce small mean deviation errors from exact values. This procedure warrants large savings in the total computing time consumed without loss of accuracy. It has been shown that the so-called PSA (Primitive basis functions, s-type functions, Analytic integration) and BSA (Basis functions, s-type functions, Analytic integration) methods are the most suitable when performing MQSM from fitted densities. In the PSA method the Gaussian function set $\Gamma(r)$ has been chosen to be the same as the squared molecular s-type renormalized primitive functions. Linear dependency problems (as those due to the presence of sp shells) have been avoided by using a set of independent set of Gaussian functions. For example, if one uses the 6-31G* [17] basis set, each H atom contributes with four Gaussian functions centered at its nuclei in the expansion of Eq. (4), whereas each C, O or F add 11 Gaussian functions to the expansion or each S atom increases in 17 Gaussian functions the total number used in the expansion. In the BSA method the set of $\{g_i\}$ Gaussian functions has been chosen to be the square s-type renormalized basis functions, the main difference with the PSA method being the presence, in the BSA case, of crossed terms among primitive functions. In both methods, integrations are performed analytically over all the space. Optimal numbers of functions to represent the fitting are currently being studied.

The SCF wavefunctions from which the electron density is fitted was calculated by means of the GAUSSIAN-90 system of programs [18]. The program QMOLSIM [3] used in the computation of the MQSM allows optimization of the mutual orientation of the two systems studied in order to maximize their similarity by the common steepest-descent, Newton and quasi-Newton algorithms [19]. The DIIS procedure [20] has been also implemented for the steepest-descent optimizations in order to improve the performance of this method. The MQSM used in the optimization procedure are obtained from fitted densities. This speeds the process. The exact MQSM were obtained from the molecular orientation obtained in this optimization procedure.

5 Applications of MQSM

The goal of the discussion in the next sections is to show several examples of how fitted densities can be used in order to study large systems where their size prevents the calculation of exact MQSM. In the first section, a study of molecular properties ordering is discussed. Further, calculations of MQSM obtained from fitted densities are applied to the prediction of the activity for a series of metal-substituted enzyme models. Finally, the use of these MQSM as an interpretative tool in chemical reactivity is discussed.

5.1 Fluoro- and Chloro-Substituted Methanes: Ordering and Physical Properties

Table 1 presents a list of the boiling and melting points [21] of nine fluoro- and chloro-methanes. We propose this set of molecules as a standard in order to compare MQSM obtained from different methodologies. The wavefunction and geometry of the molecules have been obtained by means of the GAUSSIAN-90 program [18]. Full geometry optimization of the nine molecules have been carried out using the 6-31G** [17] basis set. A Similarity study has been performed over this molecular set in order to correlate the molecules with respect to the mentioned properties. Tables 2–4 contain the fitted MQSM obtained for all the possible pairs of molecules. These measures are Overlap-like, Coulomb-like and Gravitational-like MQSM, respectively. Tables 5–7 contain the respective exact MQSM.

Figure 1 shows a Kruskal tree obtained from the Overlap-like MQSM. The dashed lines are the links of the tree. The molecular set has been divided into two classes represented by squares and rhombuses. The same figure has been obtained for both properties and it can be seen how an order has been established between the molecules: CCl_4, $CHCl_3$, CH_2Cl_2, CH_3Cl, CH_3F, CH_2F_2, CHF_3, CF_4. The methane molecule does not have a fixed position in this sequence but lies in some place at the end of the sequence and near to the CH_3F molecule. The same order for the molecules is found when dealing with other types of MQSM. For example, Fig. 2 shows the Kruskal tree related to a Coulomb-like MQSM. There the classes are defined by means of the melting points of the molecules. In this case, the measure values have been obtained from fitted densities.

In Fig. 3 the tree obtained from gravitational MQSM is shown. In Figs. 4 and 5 it can be seen how the Carbó Index applied over the gravitational MQSM correlates the molecules both with respect to the boiling and the melting points respectively. In Fig. 4 it can be seen how the group of molecules (CH_3Cl, CH_2F_2, CH_3F) acts as a bridge between the classes having the extreme

Table 1. Boiling and melting points of nine fluoro- and chloro-methanes

Molecule	Melting point (°C)[a]	Boiling point (°C)[a]
CH_4	− 182.0	− 164.0
CF_4	− 150.0	− 129.0
CHF_3	− 160.0	− 82.2
CH_3F	− 141.8	− 78.4
CH_2F_2	− 136.0	− 51.6
CH_3Cl	− 97.1	− 24.2
CH_2Cl_2	− 95.1	40.0
$CHCl_3$	− 63.5	61.7
CCl_4	− 23.0	76.5

[a] Data from [21]

Table 2. Fitted overlap-like MQSM for the fluoro- and chloro-methanes

	CH₄	CH₃F	CH₃Cl	CH₂F₂	CH₂Cl₂	CHF₃	CHCl₃	CF₄	CCl₄
CH₄	31.8371	28.6490	22.7665	28.5887	23.5142	30.7261	28.0163	35.1939	32.9459
CH₃F		151.071	316.733	147.149	314.154	145.034	309.151	145.808	302.347
CH₃Cl			1028.12	140.203	1021.52	316.281	1020.74	316.528	1020.89
CH₂F₂				270.377	287.801	255.606	280.037	247.833	269.939
CH₂Cl₂					2024.46	150.553	1734.51	288.213	1362.31
CHF₃						389.711	53.6961	385.436	54.4981
CHCl₃							3020.86	43.9439	2689.54
CF₄								509.002	64.6780
CCl₄									4017.26

Table 3. Fitted coulomb-like MQSM for the fluoro- and chloro-methanes

	CH₄	CH₃F	CH₃Cl	CH₂F₂	CH₂Cl₂	CHF₃	CHCl₃	CF₄	CCl₄
CH₄	65.7662	96.3662	113.250	127.806	160.784	159.719	207.727	191.623	253.901
CH₃F		213.356	319.085	259.842	385.188	307.278	450.621	354.732	516.195
CH₃Cl			547.313	338.688	639.100	446.752	731.516	506.866	823.865
CH₂F₂				392.349	524.552	454.535	619.431	517.077	713.499
CH₂Cl₂					1117.12	444.132	1249.02	698.024	1377.29
CHF₃						603.695	783.878	682.084	909.341
CHCl₃							1777.86	895.391	1953.93
CF₄								846.075	1120.22
CCl₄									2531.29

Table 4. Fitted gravitational-like MQSM for the fluoro- and chloro-methanes

	CH$_4$	CH$_3$F	CH$_3$Cl	CH$_2$F$_2$	CH$_2$Cl$_2$	CHF$_3$	CHCl$_3$	CF$_4$	CCl$_4$
CH$_4$	131.044								
CH$_3$F	144.316	486.657							
CH$_3$Cl	135.926	836.551	2030.72						
CH$_2$F$_2$	162.877	503.821	544.850	851.086					
CH$_2$Cl$_2$	155.896	852.246	2048.95	925.286	3945.99				
CHF$_3$	185.309	523.765	889.731	863.515	669.758	1224.41			
CHCl$_3$	181.331	868.822	2077.00	947.799	3816.93	700.257	5877.51		
CF$_4$	210.825	548.316	913.544	883.443	979.938	1251.78	715.481	1605.50	
CCl$_4$	205.870	883.715	2107.34	968.146	3601.71	743.912	5732.53	896.223	7825.90

Table 5. Exact overlap-like MQSM fot the fluoro- and chloro-mehtanes

	CH$_4$	CH$_3$F	CH$_3$Cl	CH$_2$F$_2$	CH$_2$Cl$_2$	CHF$_3$	CHCl$_3$	CF$_4$	CCl$_4$
CH$_4$	31.8387								
CH$_3$F	28.5322	151.096							
CH$_3$Cl	22.6761	316.583	1028.14						
CH$_2$F$_2$	28.3636	147.022	140.065	270.397					
CH$_2$Cl$_2$	23.3821	314.042	1021.42	287.425	2024.50				
CHF$_3$	30.3907	144.766	316.165	255.437	150.146	389.709			
CHCl$_3$	27.8498	308.990	1020.54	279.603	1734.33	52.5337	3020.92		
CF$_4$	34.7206	145.386	316.315	247.473	287.793	385.233	42.6669	508.974	
CCl$_4$	32.7276	302.104	1020.58	269.375	1361.83	52.9746	2689.20	62.7056	4017.35

Table 6. Exact coulomb-like MQSM for the fluoro- and chloro-methanes

	CH_4	CH_3F	CH_3Cl	CH_2F_2	CH_2Cl_2	CHF_3	$CHCl_3$	CF_4	CCl_4
CH_4	65.3521	95.5752	112.369	126.606	159.624	158.131	206.382	189.671	252.373
CH_3F		212.882	318.339	258.566	384.019	305.244	449.001	351.967	513.990
CH_3Cl			546.593	337.317	637.803	444.697	729.769	504.092	821.673
CH_2F_2				391.030	523.173	452.017	617.392	513.394	710.547
CH_2Cl_2					1116.31	441.963	1247.73	694.726	1375.45
CHF_3						600.833	780.791	677.639	904.580
$CHCl_3$							1777.24	890.458	1952.81
CF_4								840.997	1114.28
CCl_4									2531.08

Table 7. Exact gravitational-like MQSM for the fluoro- and chloro-methanes

	CH_4	CH_3F	CH_3Cl	CH_2F_2	CH_2Cl_2	CHF_3	$CHCl_3$	CF_4	CCl_4
CH_4	130.746	143.047	134.545	160.719	153.907	182.350	178.935	207.070	203.045
CH_3F		486.419	835.327	502.104	851.037	520.744	867.148	544.053	881.183
CH_3Cl			2030.36	543.149	2047.19	887.475	2074.11	910.285	2103.32
CH_2F_2				850.264	923.131	860.778	944.979	878.906	963.991
CH_2Cl_2					3945.86	666.805	3815.41	976.270	3598.63
CHF_3						1222.52	693.395	1247.65	733.793
$CHCl_3$							5877.86	706.294	5731.29
CF_4								1602.20	883.448
CCl_4									7826.87

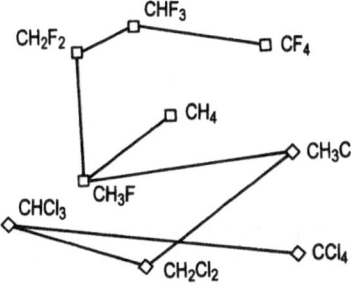

Fig. 1. Kruskal tree computed from Euclidean distances evaluated with overlap-like similarity measures for fluoro- and chloro-methanes. The classes identify two ranges of boiling or melting points

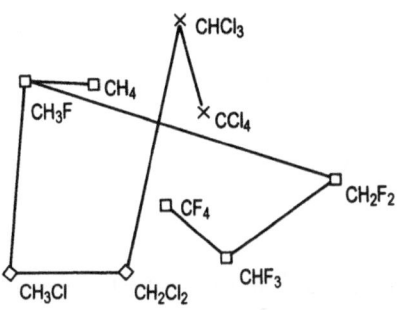

Fig. 2. Kruskal tree computed from Euclidean distances evaluated with Coulomb-like similarity measures. The classes identify two ranges of melting points

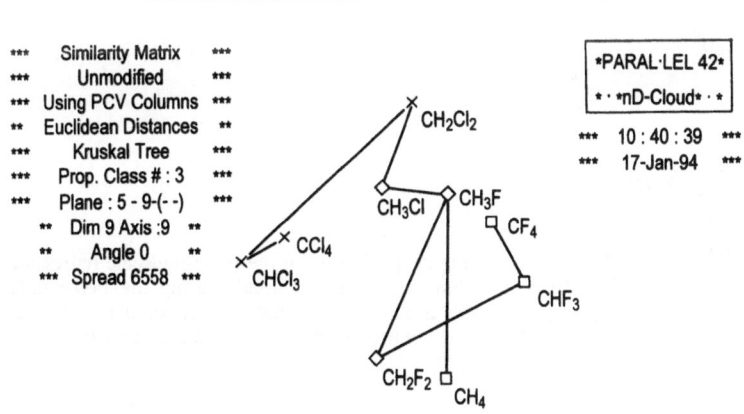

Fig. 3. Kruskal tree computed from Euclidean distances evaluated with gravitational-like similarity measures. The studied property is the boiling point

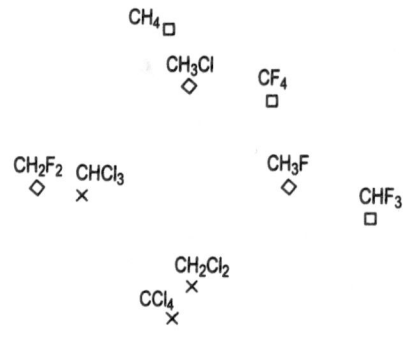

Fig. 4. Projection of the point-molecule representation of the nine molecules using the Carbó index obtained from a gravitational-like similarity measure. The studied property is the boiling point

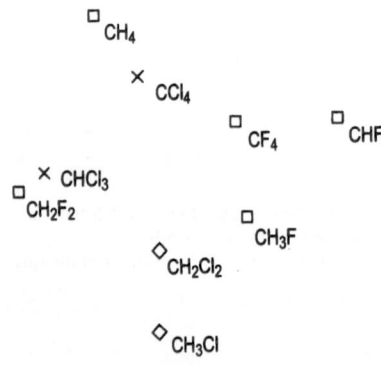

Fig. 5. Projection of the point-molecule representation of the nine molecules using the Carbó index obtained from a gravitational-like similarity measure. The studied property is the melting point.

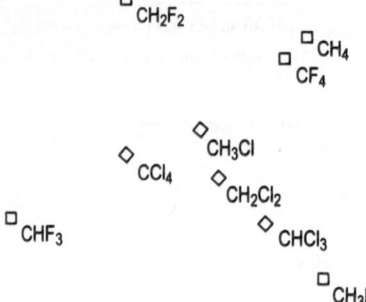

Fig. 6. Projection of the point-molecule representation of the nine molecules using the Petke index obtained from a gravitational-like similarity measure. The studied properties are boiling and melting points

range of values of the boiling point. The same study has been done using the Petke Index as it is shown in Fig. 6. There, the Chloro-methanes are surrounded by the Fluoro-methanes. The same figure is obtained when dealing with melting points.

5.2 Metal-Substituted Carbonic Anhydrases: A Successful Case

5.2.1 Introduction

Carbonic Anhydrase (CA) is a zinc metalloenzyme found both in animals and plants. Its biological function is to catalyze the reversible hydration of CO_2 and dehydration of the bicarbonate anion very efficiently [22, 23]. X-ray data [22e, 24] show that the zinc atom placed at the active site of CA is bound to three imidazole groups coming from His94, His96, and His119, and a water molecule completes a nearly symmetrical tetrahedral coordination geometry.

Substitution of the native zinc in CA by another metal has been used to shed light on the nature of the mechanism acting in this catalytic process. Zinc(II) is a d^{10} metal ion with no d-d transitions, with no unpaired electrons, and with nuclear properties which cannot be explored by NMR. Therefore, to gain more insight into the structure of CA, the naturally occurring zinc is usually replaced by high-spin Co(II) (CoCA) [25], which is recognized as a powerful spectroscopic probe is biological systems [26]. In this case, it is presumed that the structural properties of CoCA deduced from spectra can be transferred to the native enzyme. In fact, it is well known that the CA active site binds, among other dications, zinc, cobalt, copper, nickel, manganese, cadmium, and mercury [22a, 27–30]. Nevertheless, it has been found that only Co(II) and Zn(II) confer appreciable activity to the enzyme [22a, 30a]. Other metal-substituted CAs do not exhibit any substantial biological activity.

Our purpose is to make a quantitative comparison between different metal-substituted CAs to know whether the mere structural and electronic properties of a first-shell model of the CA active site can be used to explain the different behaviour of the metal-substituted CA. In particular, MQSM will be performed in order to classify a series of eight dications which have been used in metal-substituted experimental and theoretical models of CA. These eight dications are Be(II), Mg(II), Mn(II), Co(II), Ni(II), Cui(II), Zn(II) and Cd(II).

5.2.2 Methodology

Owing to computational limitations, the model used to represent the first shell of CA was the $(NH_3)_3M^{II}(H_2O)$ system. This complex has been widely used in ab initio calculations of different mechanistic aspects of the CA enzyme [31]. Furthermore, the validity of this model, where the three imidazole ligands are substituted by three ammonia groups, has been previously assessed [31e, 32]. In the present study, full geometry optimizations of all eight minima with no symmetry constraints were performed. Pseudopotential wave functions have been preferred to full-electron calculations, because valence Similarity can be more easily related to reactivity than all-electron Similarity [33]. Dunning's valence double zeta quality basis set [34] together with the Hay and Wadt ECPs

49

[35] to represent the core electrons (basis set labelled *lanl1dz* in the GAUS-SIAN-90 program [18]) were used in all calculations. The fitting procedure was the aforementioned BSA method.

5.2.3 Results and Discussion

The main structural parameters of the fully optimized geometries of the eight $(NH_3)_3M^{II}(H_2O)$ complexes studied are given in Table 8. Table 9 gathers the Euclidean distances calculated using Eq. (25) between the eight metal-substituted models of CA.

Although there are no experimental data on Be(II)-substituted models of CA, models with Be(II) instead of Zn(II) have been used in several theoretical works to study the mechanism of action of CA, given that the small size of this dication permits calculations to be performed at higher levels of theory [36]. Despite ZnCA being the closest to BeCA, the large distance between these two models of CA strongly indicates that this dication should not be used in models of CA [37].

To our knowledge, no Mg(II)-substituted CA has yet been synthesized as a model of CA, although the intermediate distance between MgCA and CoCA

Table 8. Distances from the metal to the oxygen atom (d_{M-O}) and to the nitrogen atom (d_{M-N}), and $\angle OMN$ angle for the optimized geometries of the eight $(NH_3)_3M^{II}(H_2O)$ complexes studied. Distances are given in Å and angles in degrees

Metal (M)	d_{M-O}	d_{M-N}	$\angle OMN$
Be(II)	1.712	1.761	106.7
Mg(II)	2.017	2.140	108.2
Mn(II)	2.205	2.292	107.3
Co(II)	2.101	2.174	106.2
Ni(II)	2.099	2.135	102.1
Cu(II)	2.092	2.123	131.1
Zn(II)	2.071	2.107	105.5
Cd(II)	2.277	2.294	104.9

Table 9. Distance matrix for the studied metal-substituted models of CA obtained from Eq. (25)

	Be(II)	Mg(II)	Mn(II)	Co(II)	Ni(II)	Cu(II)	Zn(II)	Cd(II)
Be(II)	0.000							
Mg(II)	11.949	0.000						
Mn(II)	13.373	7.887	0.000					
Co(II)	12.046	5.450	6.149	0.000				
Ni(II)	12.139	9.250	9.595	7.437	0.000			
Cu(II)	13.379	13.824	13.117	12.394	12.828	0.000		
Zn(II)	11.456	7.374	9.108	4.691	7.455	12.051	0.000	
Cd(II)	13.419	9.362	3.486	6.438	9.397	12.611	8.348	0.000

or ZnCA may suggest the use of this dication in experimental models of CA. An earlier theoretical work has already pointed out the large parallelism in the energy profile of a simple model of ZnCA and MgCA [38]; furthermore, it has been shown that Zn(II) exhibits intermediate behaviour between Be(II) and Mg(II), yet closer to Mg(II). This result is well reproduced by our MQSM, the distance between BeCA to ZnCA (11.46) being larger than the distance found between MgCA and ZnCA (7.37).

The smallest distance of all the related metal-substituted models of CA are found between Mn(II) and Cd(II). Geometrical parameters of Table 8 show an almost identical geometrical structure for these two complexes. This large Similarity can be related to the electronic structure of high-spin Mn(II) and Cd(II).

The most interesting result turns out to be the small distance and the large QSI found between CoCA and ZnCA, in agreement with the experimental fact that only Co(II)-substituted CA confers meaningful activity to the CA enzyme [30a]. This large similarity must be attributed to both structural and electronic resemblances. First, and as previously reported [32, 39], the geometrical parameters of Zn(II)- and Co(II)-substituted CA are quite similar (see also Table 8). Further, the electronic structure of these two cations, d^7 for high-spin Co(II) and d^{10} for Zn(II), gives rise to comparable electronic behaviour in tetrahedral geometries.

Ni(II) and Cu(II) exhibit large dissimilarities when compared to all other dications. This result can be rationalized taking into account the well-known Jahn-Teller distortion suffered by these two cations in tetrahedral complexes.

The remarkable distance found between the Cd(II)-substituted model of CA and the Zn(II) model must be attributed to the different ionic radii of these dications, given that the electronic properties of these two dications are quite analogous.

The above discussion has focused mainly on the reasons accounting for the differences and similarities between the studied metal-substituted CA. However, a global comparison with experiment is still missing. For such purpose, we must refer to the paper by Coleman [30c] which carries out an experimental study on the kinetics of metal-substituted CA. Thus, in Table 10 we collect the Euclidean distances from a given metal-substituted CA to zinc model of CA, together with the experimental activity of metal-substituted CA referred to the hydration of CO_2 catalyzed by different metal-substituted CA. It is remarkable the large correlation found between activities and Euclidean MQSM distances: the larger the distance, the smaller the experimental activity of CA. Not only the experimental ordering is correctly reproduced, but also a semiquantitative value for activities is reflected. Moreover, Table 10 can allow us to anticipate the experimental activities of Mg(II)- and Be(II)-substituted CA. The MQSM distances of such enzymes to the model native enzyme are 7.37 and 11.56, respectively (see Table 9). If one correlates the values of Table 10, one obtains $\ln Ac = 9.58 - 0.40d$, with a correlation coefficient of 0.96, which translates into a confidence of 0.1% [40]. From this equation the value of d = 7.34 for Mg translates

Table 10. Euclidean distances referred to zinc-enzyme and experimental hydration of CO_2 activities of metal-substituted enzymes

Metal	d	Ac[a]
Zn(II)	0.00	10200
Co(II)	4.69	5700
Ni(II)	7.46	500
Cd(II)	8.35	430
Mn(II)	9.11	400
Cu(II)	12.05	127

[a] Activities from [30c]

into an activity of 759. Therefore, it is expected that substitution of Zn(II) by Mg(II) would not confer appreciable experimental activity to CA. Likewise, the distance from Be(II) results in an interpolated activity of 148, so Be(II) is not likely to confer activity to CA.

To sum up, it has been shown that MQSM can be very useful when one tries to classify a series of metal-substituted enzymes as a function of their biological activities. In particular, for the CA enzyme, it has been possible to predict theoretically, with a simple model of CA and the use of MQSM, that Co(II) is the only dication that can induce appreciable activity to a metal-substituted Carbonic Anhydrase. The small Euclidean distance between CoCA and ZnCA has been associated with their small structural and electronic differences.

5.3 An Insight into Chemical Reactivity

5.3.1 The Hammond Postulate

The Hammond postulate [41] is a qualitative assumption that interrelates structural similarities between reactants, transition states, and products with the endo- or exothermicity of chemical reactions. It states that if the transition state is near in the potential energy surface to an adjacent stable complex, then it is also near in structure to the same complex. This postulate, which can be applied to most chemical reactions, has been used for predicting the effects of substituent changes and external perturbations on transition-state geometry [42, 43]. It is basically accomplished if slopes and matrices of force constants associated with reactants and products are similar [44, 45]. As yet, the unique attempt to quantify the Hammond postulate within a quantum mechanical framework is due to Cioslowski [46]. He has shown that the quantification of the Hammond postulate can be achieved by defining two new parameters, the isosynchronicity (α) parameters between transition state and stable complexes and the so-called structural proximity of the transition state to the reactans (β). These α and β definitions hold for any expression for the distance between A and B ($d_{A,B}$)

which satisfies the classical properties of a Euclidean distance. In particular, α is given by:

$$\alpha = [d_{A, TS} + d_{B, TS}]/d_{A, B}. \tag{44}$$

This parameter is never smaller than 1, and takes a value near to 1 when either the distance of A or B to the TS is small, or, in the case of reactions, where the transition state loses its similarity to the reactants at exactly the same degree as it gains its similarity to the products.

Furthermore, Cioslowski defined the parameter β as:

$$\beta = [d_{A, TS} - d_{B, TS}]/d_{A, B} \tag{45}$$

taking values from -1 to 1. If the reactants (A) and transition state (TS) are closer in the potential energy surface than products (B) and TS, then the β value is negative, and positive otherwise. So a negative value of β correspond to a reaction with an "early" TS, and a positive value to a reaction with a "late" TS.

Our interest is in using the α and β parameters together with the MQSM and distance defined by Eqs. (14) and (25) as an interpretative tool to understand better how a series of reactions proceed along the reaction coordinate, and to analyze the changes undergone by the molecular systems when going from reactants to products. This methodology was applied to the study of the $F_2S_2/FSSF$, HNC/HCN, $H_2SO/HSOH$ and $H_2SCH_2/HSCH_3$ gas-phase rearrangement reactions which have both Hammond and anti-Hammond behaviours.

5.3.2 Methodology

The 6-31G* [17] basis set of Pople et al. was used in all calculations presented in this section. The SCF wavefunctions from which the electron density is fitted were calculated by means of the GAUSSIAN-90 [18] system of programs. Full optimization of the structure of minima and transition states has been performed. For the computation of the Intrinsic Reaction Coordinate (IRC) the algorithm of Gonzalez and Schlegel [47] was followed with mass-weighted coordinates in order to get a net reaction coordinate physically meaningful. The fitting method used in this work has been the aforementioned PSA method.

5.3.3 Results and Discussion

Table 11 gathers the relative energy of reactants (A), products (B) and transition states (TS) of these four well studied rearrangement reactions (see [48] for $F_2S_2/FSSF$, [49] for HNC/HCN [50] for $H_2SO/HSOH$ and [50, 51] for $H_2SCH_2/HSCH_3$). In this table reactants have been selected to be the nearest point in energy to the transition states. In Table 12 the QSI, distances, and α and β parameters are given.

From the energy values of Table 11 and in the light of the Hammond postulate, it can be expected that the transition state of reaction $F_2S_2/FSSF$ will

Table 11. Relative energies (in kcal/mol) referred to reactants for the different species involved in the F_2S_2/FSSF, HNC/HCN, H_2SO/HSOH and H_2SCH_2/HSCH$_3$ rearrangement reactions obtained with the 6-31G* basis set

| | ΔE | | | |
	F_2S_2/FSSF	HNC/HCN	H_2SO/HSOH	H_2SCH_2/HSCH$_3$
A	0.0	0.0	0.0	0.0
TS	72.5	39.8	53.8	8.0
B	− 6.3	− 12.5	− 32.4	− 95.2

Table 12. Similarity Indices (SI), distances (d), structural proximity (β) and isosynchronicity (α) parameters for the F_2S_2/FSSF, HNC/HCN, H_2SO/HSOH and H_2SCH_2/HSCH$_3$ series of reactions

Reactions	$SI_{A,B}$	$SI_{A,TS}$	$SI_{B,TS}$	$d_{A,B}$	$d_{A,TS}$	$d_{B,TS}$	α	β
F_2S_2/FSSF	0.9257	0.9405	0.9188	14.2944	12.8092	14.9183	1.9397	− 0.1476
HNC/HCN	0.9969	0.9976	0.9968	0.9140	0.8050	0.9373	1.9062	− 0.1447
H_2SO/HSOH	0.9816	0.9783	0.9983	4.4889	4.8788	1.3593	1.3897	0.7840
H_2SCH_2/HSCH$_3$	0.9834	0.9774	0.9675	4.0878	4.7313	5.6990	2.5516	− 0.2367

be closer to F_2S_2 than to FSSF. Effectively, from the QSI and distances of Table 12 corresponding to this particular reaction, it can be seen that this result is properly reproduced: the distance of the TS to F_2S_2 (12.81) is shorter than that of the TS to the FSSF (14.92), thus giving rise to a negative value of β in this reaction.

As regards the HNC/HCN rearrangement reaction, the most noticeable fact is the small distances found between reactant, TS and product. In this reaction the C–N bond length is almost unchanged when going from reactant to product, thus explaining the large values of the Similarity Indices and the small distances between stationary points.

Nevertheless, the most interesting rearrangement reaction turns out to be the H_2SO/HSOH reaction. For this reaction the IRC has been depicted in Fig. 7. From this figure and from the values of the net reaction coordinate for reactants (− 2.69) and products (2.19) it becomes apparent that this rearrangement reaction violates the predictions of the Hammond postulate. In this case the TS is structurally closer to HSOH than to H_2SO, despite the fact that energetically this TS is closer to H_2SO than to HSOH. MQSM, as defined here, reproduce this anti-Hammond behaviour, the distance between reactant and TS being larger (4.88) than the distance between product and TS (1.36). The positive value of β also shows that the TS is closer to product than to reactant. Finally, the small value of α due to the small distance between product and TS, reflects the large, structurally speaking, similarity between product and TS. From Fig. 7 it can be clearly seen that the force constant associated with reactants and products in the direction given by the transition vector are quite different, the

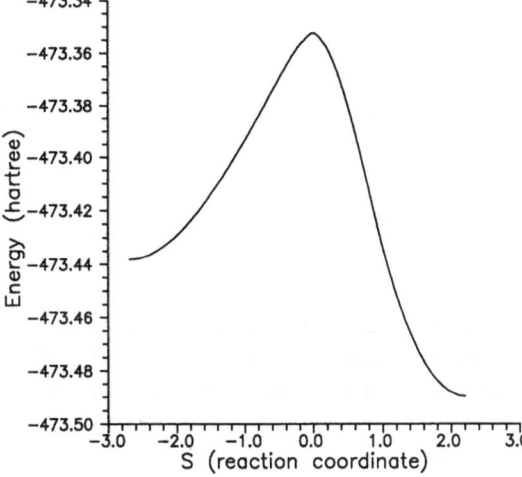

Fig. 7. Intrinsic Reaction Path of $H_2SO/HSOH$ rearrangement reaction computed with the 6-31G* basis set

force constant associated with products being larger than that associated to reactants, and therefore accounting for the anti-Hammond behaviour. Along the reaction coordinate, the main change in this reaction is the breaking of a S–H bond and the formation of a new O–H bond. The different force constant associated with reactants and products reflects essentially the fact that the O–H bond in HSOH is stronger than the S–H bond in H_2SO.

The $H_2SCH_2/HSCH_3$ rearrangement reaction is also found to follow Hammond behaviour, the TS being closer to H_2SCH_2 than to $HSCH_3$. The distance between TS and H_2SCH_2 is found to be 4.73, whereas that for TS and $HSCH_3$ is 5.70. The negative value of β reflects again the Hammond behaviour; the TS is more similar to the reactant than to the product. Interestingly, the large value of α is a consequence of the similarity between reactant and product being larger than either the similarity between TS and reactant or product and reflects also the asymmetry of the energy profile for this reaction.

In conclusion, the previous show that, by taking advantage of the information provided by the MQSM, it is possible to draw an approximate shape of the computationally very demanding Intrinsic Reaction Coordinate, independent of the definition of the reaction coordinate being more or less complicated, and also allowing the prediction of the Hammond and anti-Hammond behaviour.

Acknowledgments. This work has been financed by the CICYT-CIRIT, Fine Chemicals Programme of the "Generalitat de Catalunya" through grant: #QFN91-4206 and by the spanish DGICYT project #PB92-0333. One of us (E.B.) benefits from a grant from the "Department d'Ensenyament de la Generalitat de Catalunya".

E. Besalú et al.

Appendix: Practical Implementation of Gravitational Integrals

Gaussian Functions Definition

Following the notation of Obara and Saika [52] we can define an unnormalized Cartesian Gaussian Function with origin at $R = (R_x, R_y, R_z)$, with orbital exponent ζ and quantum numbers collected in the vector $n = (n_x, n_y, n_z)$ as:

$$\varphi(r, \zeta, n, R) = (x - R_x)^{n_x}(y - R_y)^{n_y}(z - R_z)^{n_z}\, e^{-\zeta|r - R|^2}, \tag{A1}$$

where the vector $r = (x, y, z)$ contains the electron Cartesian coordinates.

An s-type function bears the quantum numbers vector $n = 0 = (0, 0, 0)$ while a p-type function has the quantum numbers collected in the vector $1_i = (\delta_{ix}, \delta_{iy}, \delta_{iz})$ which represents the p_x, p_y and p_z components when $i = x, y, z$ respectively. Higher order angular terms can be obtained by means of a combination of the previously defined ones. For example, the six terms of a d shell are obtained from the expresson $1_i + 1_j$ $(i, j = x, y, z)$.

We also define the function which returns the value of the i-th component of a vector n:

$$N_i(n) = n_i. \tag{A2}$$

The Gravitational Integral

We define the Gravitational Integral between four functions as outlined in Eq. (21). When AO are expressed in terms of Gaussian functions, the integral reads:

$$\langle ab|r_{12}^{-2}|cd\rangle = (ab, cd) = \int \varphi(r_1, \zeta_a, a, A)\,\varphi(r_1, \zeta_b, b, B)$$
$$\times |r_1 - r_2|^{-2}\,\varphi(r_2, \zeta_c, c, C)\,\varphi(r_2, \zeta_d, d, D)\,dr_1\,dr_2. \tag{A3}$$

Using the Gaussian Transform:

$$|r_1 - r_2|^{-2} = \int_0^\infty e^{-|r_1 - r_2|^2 u}\,du, \tag{A4}$$

we obtain

$$(ab, cd) = \int_0^\infty (ab|u|cd)\,du, \tag{A5}$$

where

$$(ab|u|cd) = \int \varphi(r_2, \zeta_c, c, C)\,\varphi(r_2, \zeta_d, d, C)\,(a|0_{r_2}|b)\,dr_2, \tag{A6}$$

and the three center overlap integral has been used:

$$(a|0_{r_2}|b) = \int \varphi(r_1, \zeta_a, a, A)\,\varphi(r_1, u, 0, r_2)\,\varphi(r_1, \zeta_b, b, B)\,dr_1. \tag{A7}$$

Note that the transform coming from Eq. (A4) creates, within the integral at Eq. (A7), an s-type function centered at r_2 with exponential parameter u.
Defining:

$$(ab, cd)^{(m)} = \int_0^\infty \left(\frac{u}{\rho + u}\right)^m (ab|u|cd)\, du, \tag{A8}$$

and following the notation of Obara and Saika, the recursive formula is found:

$$((a + 1_i)b, cd)^{(m)} = (P_i - A_i)(ab, cd)^{(m)} + (W_i - P_i)(ab, cd)^{(m+1)}$$

$$+ \frac{1}{2\zeta} N_i(a) \left\{ ((a - 1_i)b, cd)^{(m)} - \frac{\rho}{\zeta}((a - 1_i)b, cd)^{(m+1)} \right\}$$

$$+ \frac{1}{2\zeta} N_i(b) \left\{ (a(b - 1_i), cd)^{(m)} - \frac{\rho}{\zeta}(a(b - 1_i), cd)^{(m+1)} \right\}$$

$$+ \frac{1}{2(\zeta + \eta)} N_i(c)(ab, (c - 1_i)d)^{(m+1)}$$

$$+ \frac{1}{2(\zeta + \eta)} N_i(d)(ab, c(d - 1_i))^{(m+1)} \tag{A9}$$

which holds for $i = x, y, z$ and, for $m = 0$, stands for a true Gravitational Integral as the one defined in Eq. (A3). In Eq. (A9) the following definitions have been taken into account:

$$\zeta = \zeta_a + \zeta_b, \quad \eta = \zeta_c + \zeta_d, \tag{A10}$$

$$P = \frac{\zeta_a A + \zeta_b B}{\zeta}, \quad Q = \frac{\zeta_c C + \zeta_d D}{\eta}, \quad W = \frac{\zeta P + \eta Q}{\zeta + \eta} \tag{A11}$$

and

$$\rho = \frac{\zeta \eta}{\zeta + \eta}. \tag{A12}$$

Use of the recurrence formula at Eq. (A9) requires the explicit evaluation of the following integral:

$$(0_A 0_B, 0_C 0_D)^{(m)} = S_{AB} S_{CD} \int_0^\infty \left(\frac{u}{\rho + u}\right)^m \left(\frac{\rho}{\rho + u}\right)^{3/2} e^{-\frac{\rho u}{\rho + u}|PQ|^2}\, du \tag{A13}$$

where S_{AB} stands for an unnormalized overlap integral between s-type functions centered in A and B. The respective definition holds for S_{CD}.
Performing the variable transform:

$$\frac{u}{\rho + u} = 1 - t^2, \tag{A14}$$

one finally gets:

$$(0_A 0_B, 0_C 0_D)^{(m)} = 2\rho\, S_{AB} S_{CD}\, G_m(\rho|PQ|^2), \tag{A15}$$

where the auxiliary integral

$$G_m(T) = \int\limits_0^1 (1 - t^2)^m e^{-T(1-t^2)} dt \qquad (A16)$$

is defined.

The discussion above shows how to compute the Gravitational Integral through a recursive path, as well as that a gravitational integral can be obtained from any formula standing for an electron repulsion integral which is computed by means of the well known $F_m(T)$ integral set: it is only necessary to replace the $F_m(\rho|PQ|^2)$ terms by the $(\pi\rho)^{1/2} G_m(\rho|PQ|^2)$ ones.

Evaluation of the $G_m(T)$ Integral

From [53] it can be deduced that:

$$G_m(T) = G_m(0) e^{-T} M(\tfrac{1}{2}, \tfrac{3}{2} + m, T) \qquad (A17)$$

where

$$G_m(0) = \frac{(2m)!!}{(2m + 1)!!} \qquad (A18)$$

and $M(\tfrac{1}{2}, \tfrac{3}{2} + m, T)$ is the Kummer Hypergeometric confluent function [53].

As far as we know, the evaluation of the $G_m(T)$ integrals presents as many difficulties as the evaluation of the classical $F_m(T)$ ones. There are many numerical methods which lead to the evaluation of the integral. Some of them are listed below.

There are series which can be used to obtain the values of the Kummer functions but they are only numerically stable for a concrete range of the argument T.

As:

$$\frac{dG_m(T)}{dT} = -G_{m+1}(T), \qquad (A19)$$

it is also feasible to think about a tabulation of the integral for a set of arguments and then interpolate, following the rule:

$$G_m(T) = \sum_{i=0}^{\infty} \frac{G_{m+1}(T_o)}{i!}(T_o - T)^i. \qquad (A20)$$

This possibility has not been explored yet in our laboratory but, for small values of the argument T, it is possible to apply Eq. (A20) with $T_o = 0$ combined with Eq. (A18).

There is another choice consisting of applying the binomial expansion in Eq. (A16) to

$$G_m(T) = e^{-T} \sum_{k=0}^{m} \binom{m}{k}(-1)^k Z_k(T), \qquad (A21)$$

where

$$Z_k(T) = \int_0^1 x^{2k} e^{Tx^2} dx. \tag{A22}$$

The $Z_k(T)$ integral can be computed by means of the following iterative formula:

$$Z_k(T) = (e^{-T} - 2TZ_{k+1}(T))/(2k + 1), \tag{A23}$$

which is valid when going from large to small values of the k parameter. This needs the computation of the last term of the set as:

$$Z_k(T) = \sum_{i=0}^{\infty} T^i/([2(k + i) + 1]i!). \tag{A24}$$

Another choice consists of rewriting the integral after the change of variable $1 - x^2 = t$:

$$G_m(T) = \frac{1}{2} \int_0^1 x^m (1 - x)^{-1/2} e^{-Tx} dx, \tag{A25}$$

when, one can start from Eq. (A25) and use the series:

$$(1 - x)^{-1/2} = \sum_{i=0}^{\infty} \frac{(2i - 1)!!}{(2i)!!} x^i, \quad 0 \le x < 1. \tag{A26}$$

In this manner the integral becomes:

$$G_m(T) = \frac{1}{2} \sum_{i=0}^{\infty} \frac{(2i - 1)!!}{(2i)!!} Y_{m+i}(T), \tag{A27}$$

where

$$Y_m(T) = \int_0^1 x^m e^{-Tx} dx. \tag{A28}$$

The $Y_m(T)$ integral can be computed with an iterative formula which is stable for large values of the argument T:

$$Y_m(T) = (mY_{m-1}(T) - e^{-T})/T. \tag{A29}$$

Moreover, the $Y_m(T)$ integral can be written as:

$$Y_m(T) = \frac{m!}{T^{m+1}} - A_m(T), \tag{A30}$$

where $A_m(T)$ is defined as:

$$A_m(T) = \int_1^{\infty} x^m e^{-Tx} dx, \tag{A31}$$

a well known function arising from the evaluation of integrals involving Exponential Type Orbitals [54].

Finally, for asymptotic values of the $G_m(T)$ integral, the approximation [53]:

$$G_m(T) \approx \frac{m!}{2T^{m+1}} \sum_{i=0}^{\infty} \frac{(m+1)_i(\frac{1}{2})_i}{i!\,T^i} \tag{A32}$$

can be used as an argument. There, the terms $(a)_o = 1$ and $(a)_n = a(a+1)\ldots(a+n-1)$ are defined. This last expression can degenerate into the less expensive formula

$$G_m(T) \approx \frac{m!}{2T^{m+1}} \tag{A33}$$

in some cases.

6 References

1. (a) Carbó R, Arnau M, Leyda L (1980) Int J Quantum Chem 7: 1185; (b) Carbó R, Arnau C (1981) In: de las Heras FG, Vega S (eds) Medicinal chemistry advances. Pergamon Press, Oxford, p 85; (c) Carbó R, Domingo Ll (1987) Int J Quantum Chem 23: 517; (d) Carbó R, Calabuig B (1989) Comp Phys Commun 55: 117; (e) Carbó R, Calabuig B (1990) In: Johnson MA, Maggiora G (eds) Concepts and applications of molecular similarity. John Wiley, New York, p 147; (f) Carbó R, Calabuig B (1990) In: Proceedings del XIX Congresso Internazionale dei Chimici Teorici dei Paesi di Espressione Latina. Roma, Italy, September 10–14. Carbó R, Calabuig B (1992) J Mol Struct (Teochem),254: 517; (g) Carbó R, Calabuig B (1992) J Chem Inf Comput Sci 32: 600; (h) Carbó R, Calabuig B (1992) In: Fraga S (ed) Structure, interactions and reactivity. Elsevier Pub., Amsterdam; (i) Carbó R, Calabuig B (1992) Int J Quantum Chem 42: 1681; (j) Carbó R, Calabuig B (1992) Int J Quantum Chem 42: 1695; (k) Carbó R, Calabuig B, Besalú E, Martínez A (1992) Molecular engineering 2: 43; (l) Carbó R, Besalú E, Calabuig B, Vera V, Adv Quant Chem 25: 253 (1994)
2. Carbó R, Besalú E (1994) In: Carbó R (ed) Proceedings of the first Girona Seminar on Molecular Similarity (in press)
3. Mestres J, Solà M, Besalú E, Duran M, Carbó R (1993) QMOLSIM version 1.0, Girona, CAT.
4. (a) Löwdin PO (1955) Phys Rev 97: 1474; (b) McWeeny R (1955) Proc Roy Soc A 232: 114; (c) McWeeny R (1956) Proc Roy Soc A 235: 496; (d) McWeeny R (1959) Proc Roy Soc A 253: 242
5. Cioslowski J, Fleischmann ED (1991) J Am Chem Soc 113: 64
6. Mestres J, Solà M, Duran M, Carbó R (1994) In: Mezey PG, Carbó R (eds) Proceedings of the First Girona Seminar on Molecular Similarity. (in press)
7. (a) Hodgkin EE, Richards WG (1987) Int J Quant Chem 14: 105; (b) Good AC, Hodgkin EE, Richards WG (1992) J Chem Inf Comput Sci 32: 188
8. Tou JT, González RC (1974) In: Pattern recognition principles. Addison-Wesley, Reading, MA
9. Petke JD (1993) J Comput Chem 14: 928
10. Cioslowski J, Martinov M, Mixon ST (1993) J Phys Chem 97: 10948
11. Phillips ER (1984) In: An introduction to analysis and integration theory. Dover Pub Inc, New York
12. Srivastava MS, Carter EM (1983) In: An introduction to applied multivariate statistics. Elsevier Science Pub Co Inc, New York
13. (a) Brønsted JN (1928) Chem Rev 5: 231; (b) Hammett LP (1970) In: Physical organic chemistry. McGraw-Hill, New York; (c) Marcus RA (1964) Ann Rev Phys Chem 15: 155
14. (a) Burt C, Huxley P, Richards WG (1990) J Comp Chem 11: 1139; (b) Good AC, So S-S, Richards WG (1993) J Med Chem 36: 433
15. Rouvray DH (1990) In: Johnson MA, Maggiora GM (eds) Concepts and applications of molecular similarity. John Wiley & Sons, Inc. p 15
16. (a) Mestres J, Solà M, Duran M, Carbó R (1994) J Comp Chem 15: 1113; (b) Mestres J, Solà M, Besalú E, Duran M, Carbó R (1994) In: Carbó R (ed) Proceedings of the First Girona Seminar on Molecular Similarity. (in press)

17. (a) Hehre WJ, Ditchfield R, Pople JA (1972) J Chem Phys 56: 2257; (b) Francl MM, Pietro WJ, Hehre WJ, Binkley JS, Gordon MS, Frees DJ, Pople JA (1982) J Chem Phys 77: 3654; (c) Hariharan PC, Pople JA (1973) Theor Chim Acta 28: 213

18. Frisch MJ, Head-Gordon M, Trucks GW, Foresman JB, Schlegel HB, Raghavachari K, Binkley JS, Gonzalez C, Defrees DJ, Fox DJ, Whiteside RA, Seeger R, Melius CF, Baker J, Martin RL, Kahn LR, Stewart JJP, Topiol S, Pople JA (1990) GAUSSIAN 90, Revision H, Gaussian Inc, Pittsburgh, PA

19. (a) Scales LE (1985) In: Introduction to non-linear optimization. Springer-Verlag, New York; (b) Fletcher R (1981) In: Practical methods of optimization. Wiley, Chichester; (c) Wolfe MA (1978) In: Numerical methods for unconstrained optimization. Van Nostrand Reinhold, New York; (d) Beveridge GSG, Schechter RS (1970) In: Optimization: theory and pratice. McGraw Hill Kogakuska, Tokyo; (e) Pierre DA (1969) In: Optimization theory with applications. John Wiley, New York

20. (a) Hamilton TP, Pulay P (1986) J Chem Phys 84: 5728; (b) Pulay P (1980) Chem Phys Lett 73: 393

21. (a) Lide DL (ed.) (1991) Handbook of chemistry and physics. CRC Press, Boca Raton, FL

22. (a) Wooley P (1975) Nature 258: 677; (b) Silverman DN, Lindskog S (1980) Acc Chem Res 21: 30; (c) Pocker Y, Deits TL (1983) J Am Chem Soc 105: 980; (d) Sen AC, Tu CK, Thomas H, Wynns GC, Silverman DN (1986) In: Bertini I, Luchinat C, Maret W, Zeppezauer M (eds) Zinc enzymes. Birkhäuser, Boston, p 329; (e) Eriksson EA, Jones TA, Liljas A (1986) In: Bertini I, Luchinat C, Maret W, Zeppezauer M (eds) Zinc enzymes. Birkhäuser, Boston, p 317; (f) Mulholland AJ, Grant GH, Richards WG (1993) Protein engineering 6: 133

23. (a) Davis RP (1959) J Am Chem Soc 81: 5674; (b) Pocker Y, Janjić N (1989) J Am Chem Soc 111: 731; (c) Rowlett RS, Silverman DN (1982) J Am Chem Soc 104: 6737; (d) Tu C, Silverman DN, Forsman C, Jonsson B-H, Lindskog S (1989) Biochemistry 28: 7913

24. (a) Eriksson AE, Liljas A (1993) Proteins 16: 29; (b) Kannan KK, Ramanadham M, Jones TA (1984) Ann N Y Acad Sci 429: 49; (c) Eriksson AE, Liljas A (1986) J Biol Chem 261: 16247; (d) Eriksson AE, Jones TA, Liljas A (1988) Proteins, 4: 274; (e) Kannan KK, Ramanadham M (1981) Int J Quantum Chem 20: 199

25. (a) Bertini I, Canti G, Luchinat C, Scozzafava A (1978) J Am Chem Soc 100: 4873; (b) Tu CK, Silverman DN (1986) J Am Chem Soc 108: 6065; (c) Williams TJ, Henkens RW (1985) Biochemistry 24: 2459; (d) Haffner PH, Coleman JE (1975) J Biol Chem 250: 996; (e) Pesando JM (1975) Biochemistry 14: 681–688

26. Bertini I, Luchinat C (1983) Acc Chem Res 16: 272

27. (a) Bertini I, Luchinat C, Monnani R, Roelens S, Moratal JM (1987) J Am Chem Soc 109: 7855; (b) Lanir A, Navon G (1972) Biochemistry 11: 3536; (c) Led JJ, Neesgaard E (1987) Biochemistry 26: 183

28. Tu CK, Silverman DN (1985) Biochemistry 24: 5881

29. Wilkins RG, Williams KR (1974) J Am Soc 96: 2241

30. (a) Coleman JE (1967) J Biol Chem 242: 5212; (b) Vallee BL, Gades A (1984) Adv Enzymol 56: 283; (c) Coleman JE (1967) Nature 214: 193; (d) Lipscomb WN (1983) Ann Rev Biochem 52: 17

31. (a) Jacob O, Cardenas R, Tapia O (1990) J Am Chem Soc 112: 8692; (b) Bertini I, Luchinat C, Rosi H, Sgamellotti A, Tarantelli F (1990) Inorg Chem 29: 1460; (c) Zheng Y-J, Merz KM Jr. (1992) J Am Chem Soc 114: 10498; (d) Solà M, Lledós A, Duran M, Bertrán J (1992) J Am Chem Soc 114: 869; (e) Solà M, Lledós A, Duran M, Bertrán J (1992) In: Bertrán J (ed) Molecular aspects of biotechnology: Computational models and theories. Kluwer Academic Publishers, The Netherlands, p 263; (f) Krauss M, Garmer DR (1991) J Am Chem Soc 113: 6426

32. Garmer DR, Krauss M (1992) J Am Chem Soc 114: 6487

33. Richards WG, Hodgkin EE (1988) Chemistry in Britain 1141

34. (a) Dunning TH Jr. (1970) J Chem Phys 53: 2823; (b) Dunning TH Jr. (1971) J Chem Phys 55: 3958

35. (a) Wadt WR, Hay PJ (1985) J Chem Phys 82: 284; (b) Hay PJ, Wadt WR (1985) J Chem Phys 82: 299

36. (a) Liang J-Y, Lipscomb WN (1987) Biochemistry 26: 5293; (b) Liang J-Y, Lipscomb WN (1989) Biochemistry 28: 9724; (c) Liang J-Y, Lipscomb WN (1989) Int J Quantum Chem 36: 299

37. Osman R, Weinstein H, Topiol S (1981) Ann N Y Acad Sci 367: 356

38. Solà M, Lledós A, Duran M, Bertrán J (1992) Theor Chim Acta 81: 303

39. Vedani A, Huhta DW (1990) J Am Chem Soc 112: 4759

40. Sachs L (1982) In: Applied statistics. A handbook of techniques. Springer-Verlag, New York

41. Hammond GS (1955) J Am Chem Soc 77: 334

42. (a) Duran M, Bertrán J (1990) Reports in molecular theory 1: 57; (b) Solà M, Lledós A, Duran M, Bertrán J, Abboud J-LM (1991) J Am Chem Soc 113: 2873; (c) Solà M, Carbonell E, Lledós A, Duran M, Bertrán J (1992) J Mol Struct (Theochem) 255: 283
43. (a) Formosinho SJ (1991) In: Formosinho SJ, Csizmadia IG, Arnaut LG (eds) Theoretical and computational models for organic chemistry. Kluwer Academic Publishers, Dordrecht, p 159
44. (a) Varandas AJC, Formosinho SJ (1986) J Chem Soc Faraday Trans 2 82: 953; (b) Formosinho SJ (1988) J Chem Soc Perkin Trans 2 2: 839; (c) Parr C, Johnston HS (1963) J Am Chem Soc 85: 2544; (d) Agmon N, Levine RD (1980) Isr J Chem 19: 30; (e) Agmon N (1981) Int J Chem Kinet 45: 343; (f) Agmon N, Levine RD (1977) Chem Phys Lett 52: 197; (g) Lendway G (1989) J Phys Chem 93: 4422; (h) Miller AR (1978) J Am Chem Soc 100: 1984
45. Arteca GA, Mezey PG (1988) J Comp Chem 9: 728
46. Cioslowski J (1991) J Am Chem Soc 113: 6756
47. Gonzalez C, Schlegel HB (1988) J Chem Phys 90: 2154
48. (a) Solouki B, Bock H (1977) Inorg Chem 16: 665; (b) Solà M, Schleyer PvR (to be published)
49. (a) Müller K, Brown LD (1979) Theor Chim Acta 53: 75; (b) Gray SK, Miller WH, Yamaguchi Y, Schaefer III HF (1980) J Chem Phys 73: 2733; (c) Garret BC, Redmon MJ, Steckler R, Truhlar DG, Baldridge KK, Bartol D, Schmidt MW, Gordon MS (1988) J Phys Chem 92: 1476; (d) Bentley JA, Bowman JM, Gazdy B, Lee TJ, Dateo CE (1992) Chem Phys Lett 198: 563; (e) Fan LY, Ziegler T (1992) J Am Chem Soc 114: 10890; (f) Bentley JA, Huang CN, Wyatt RE (1993) J Chem Phys 98: 5207
50. Solà M, Gonzalez C, Tonachini G, Schlegel HB (1990) Theor Chim Acta 77: 281
51. (a) Mitchell DJ, Wolfe S, Schlegel HB (1981) Can J Chem 59: 3280; (b) Eades RA, Gassman PG, Dixon DA (1981) J Am Chem Soc 103: 1066; (c) Dixon DA, Dunning Jr.TH, Eades RA, Gassman, PG (1993) J Am Chem Soc 105: 7011
52. Obara S, Saika A (1986) J Chem Phys 84: 3963
53. Abramowitz M, Stegun IA (1965) In: Handbook of mathematical functions. Dover Pub Inc, New York
54. (a) Carbó R, Besalú E (1992) Adv Quantum Chem 24: 115; (b) Carbó R, Besalú E (1992) Can J Chem 70: 353

Density Domain Bonding Topology and Molecular Similarity Measures

Paul G. Mezey

Mathematical Chemistry Research Unit, Department of Chemistry and, Department of Mathematics and Statistics, University of Saskatchewan, Saskatoon, Canada, S7N 0 W0

Table of Contents

A systematic description of chemical bonding and molecular shape is based on a simple density domain (DD) principle. An electronic density domain is the formal body enclosed by a molecular isodensity contour (MIDCO) surface $G(a)$ of some density threshold a. In the Density Domain approach, chemical bonding is described by the interfacing and mutual interpenetration of local, fuzzy charge density clouds. At high density thresholds, only disjoined, local nuclear neighborhoods appear. However, at lower density thresholds the separate density domains eventually join, and gradually change into a series of topologically different bodies. By monitoring the essential topological changes of these density domains as the density threshold a is varied, a detailed description of both the bonding pattern and the molecular shape is obtained.

Various formal molecular fragments are regarded as fuzzy moieties of electron densities, dominated by one or several nuclei. A fuzzy fragment of the electron density involves a whole range of density values and consequently cannot be described by a single MIDCO. However, such a fuzzy fragment can be represented by a sequence of density domains. There are only a *finite number of topologically different bodies of density domains* within the chemically important range of density values. As a consequence, a simple, discrete, algebraic representation is possible. Molecular shape and chemical bonding between fragments of a molecule are characterized by a finite sequence of density domains for a range of density values, by the corresponding sequence of topological patterns of their mutual interpenetration, and by the resulting algebraic structure.

1 Background and Basic Topological Concepts

The characterization of the interrelations between chemical bonding and molecular shape requires a detailed analysis of the electronic density of molecules. Chemical bonding is a quantum mechanical phenomenon, and the "shorthand" notations of formal single, double, triple, and aromatic "bonds" used by chemists are a useful but rather severe oversimplification of reality. Similarly, the classical concepts of "body" and "surface", the usual tools for the shape characterization of macroscopic objects, can be applied to molecules only indirectly. The quantum mechanical uncertainty of both electronic and nuclear positions within a molecule implies that valid descriptions of both chemical bonding and molecular shape must be based on the fuzzy, delocalized properties of electronic density distributions. These electron distributions are dominated by the nuclear arrangements and hence quantum mechanical uncertainly affects electrons on two levels: by the lesser positional uncertainty of the more massive nuclei, and by the more prominent positional uncertainty of the electrons themselves. These two factors play important roles in chemistry and affect both chemical bonding and molecular shape.

The study of electronic charge distributions [1–4] provides important theoretical insight and explanations of most molecular properties, and also serves as tool for various applied fields such as the interpretation of biochemical processes and drug design [5–7].

1.1 Molecular Isodensity Contours (MIDCOs)

According to a recent proposal [4, 8, 9], chemical bonding within formal molecular "bodies" can be described by molecular isodensity contour (MIDCO) surfaces and by density domains (DD) that are the formal bodies enclosed by such MIDCO surfaces. Originally, the DD approach has been suggested as an alternative to the conventional "skeletal model" of chemical bonding [8, 9], but the same DD approach can also be used as a tool for shape characterization. In the DD approach, the classically motivated line graphs as representatives of bonding are replaced by the pattern of interpenetration of fuzzy fragment bodies at various density thresholds. Of course, a single MIDCO surface $G(a)$ and the density domain enclosed by it cannot describe all the essential details of the electron distribution; for different electron density thresholds a the molecule exhibits a different set of MIDCOs. However, for most small changes in density threshold a, the topology of the MIDCOs does not change. If one considers an isolated molecule, then within the entire range of density thresholds a, there are only a *finite number of topologically different bodies of density domains*. Consequently, a detailed description of the fuzzy density domains of the molecule can be given not by individual MIDCOs but by *topological equivalence classes* of MIDCOs and the associated equivalence classes of density domains. This DD approach offers a detailed description of bonding within molecular "bodies", as well as a detailed shape characterization.

For a given electronic state of the molecule, the nuclear arrangement K has a dominant influence on the shape of the molecular electron distribution $\rho(K, r)$, where nuclear arrangement K is regarded as a point in the nuclear configuration space and r is the position vector in ordinary 3D space. The molecular electronic charge density function $\rho(K, r)$ can be determined experimentally, for example, from X-ray diffraction experiments [1, 2], or can be calculated by quantum chemical methods. For the calculation of electronic charge densities, standard quantum chemistry programs such as the Gaussian family of programs [10], or approximate methods designed for large molecules [11], or the more recent MEDLA approach [12] providing ab initio quality electron densities for molecules as large as proteins can be used.

The nuclear configuration space M is provided with a metric, that is, with a proper distance function $d(K, K')$ for any two nuclear arrangements K and K'. (One should keep in mind, however, that the nuclear configuration space M is not a vector space and cannot be turned into one [13].)

For the specified electronic state, the molecular potential energy is a function of the nuclear arrangements K, and can be regarded as a hypersurface $E(K)$ over this space M [13]. Each chemical species can be associated with a formal *catchment region* of the given energy hypersurface. This model relaxes the classical constraints of rigid nuclear geometry and avoids the conflict with the Heisenberg uncertainty relation [13]. In a similar manner, the electronic density of a specified electronic state can be regarded as a function $\rho(K, r)$ defined over the nuclear configuration space M. The molecular shape is represented by an entire family of $\rho(K, r)$ charge density distributions which occur within the catchment region of the molecule [4].

A formal definition of a *molecular isodensity contour surface*, MIDCO $G(K, a)$ of nuclear configuration K and density threshold a is given as the collection of points of the 3D space where the electronic density $\rho(K, r)$ is equal to the threshold value a:

$$G(K,a) = \{r: \rho(K,r) = a\}. \tag{1}$$

The molecular electronic density $\rho(K,r)$ is a continuous function of the position vector r and, consequently, the set of all points r fulfilling Eq. (1) forms a continuous surface.

Most of the electronic density is localized within a close neighborhood of the set of nuclei of the molecule. Whereas the electronic charge density function $\rho(K,r)$ becomes zero only at infinite distance from the nuclei, this function converges rapidly to zero, becoming negligible at about 8–10 Å from the nearest nucleus. In an approximate sense, we may regard only those regions of the 3D space to belong to the molecule where the electronic density $\rho(K,r)$ is larger than some small threshold. For an appropriate, small threshold value, a an approximate *molecular body* can be defined as the collection $F(K,a)$ of all those points r of the 3D space where the electronic density is greater than the value a,

$$F(K,a) = \{r: \rho(K,r) > a\}. \tag{2}$$

This set $F(K,a)$ is a *level set* of the electronic density for the threshold level a. According to the usual convection, *negative* charge means a *positive* value for the density function $\rho(K,r)$.

1.2 Density Domains (DDs)

According to the definitions by Eqs. (1) and (2), those points r of the 3D space where the value of the electronic density function is equal to the threshold

$$\rho(K,r) = a \tag{3}$$

do not belong to the level set $F(K,a)$. The closely related *density domains*, denoted by $DD(K,a)$, include *all the points and all the boundary points* of the corresponding level sets $F(K,a)$ of the same density threshold a. The density domains provide a natural representation of chemical bonding in molecules. The density domains $DD(K,a)$ are defined as

$$DD(K,a) = \{r: \rho(K,r) \geq a\}. \tag{4}$$

The shape and size of a density domain $DD(K,a)$ depend on the electronic state, the nuclear configuration K and on the choice of the threshold value a.

If one considers a conformation of an isolated molecule, then within the entire range of density thresholds a, there are only a *finite number of topologically different bodies of density domains*. According to the Density Domain Approach, chemical bonding within a molecule is described by the interfacing and mutual interpenetration of local, fuzzy charge density clouds. The bonding between molecular fragments is characterized by a sequence of density domains within a wide range of density values and by the corresponding finite sequence of topologically different patterns of the mutual interpenetration of these fragments. The DD approach is an alternative to the conventional "skeletal model" of chemical bonding: the classically motivated line diagrams of "bonds" are replaced by the pattern of interpenetration of fuzzy fragment bodies at various density thresholds.

Families of MIDCOs $G(K, a)$ and the associated density domains $DD(K, a)$ are used in the analysis of molecular shapes and molecular similarity. The Shape Group Methods (SGM) [4] are nonvisual, algebraic-topological shape analysis techniques based on the homology group of molecular MIDCO surfaces truncated according to some local curvature criteria. For the study and numerical evaluation of molecular similarity the general GSTE principle (Geometrical Similarity as Topological Equivalence) [4] is applied. The GSHAPE 90 computer program [14] evaluates the shape groups of ab initio or semiempirical electron densities, and performs nonvisual similarity evaluation of MIDCO sequences.

Since there are only a finite number of topologically different density domains for each conformation of a molecule, the bonding pattern of each conformer can be represented by a finite sequence of density domains. Most features of these sequences are specific for the given conformation of the molecule, showing some characteristic density ranges where typical shape features occur. Nevertheless, when considering the entire range of possible density threshold values, there are some important general trends that apply for most conformers of most molecules [4, 8]. These trends lead to a classification of density domains according to the ranges of the density threshold parameter a.

Starting with the high density MIDCOs near the nuclei, we consider a gradual decrease of the density threshold a. For high density thresholds a, only individual nuclear neighborhoods appear as disconnected density domains, where there is precisely one nucleus within each density domain which appears. This range of density thresholds is referred to as the *atomic range*. This atomic range can be subdivided into two subranges. In the *strictly atomic range* all the atomic density domains are convex sets, whereas in the *prebonding range* at least one density domain is no longer convex, as it "reaches out" to join eventually a neighboring density domain.

When lowering the density threshold, some nuclear neighborhoods appear to join to form density domains containing two or more nuclei. However, not all nuclei of the molecule are contained within a common density domain. Within this density range various functional groups appear as individual entities, hence this range is called the *functional group range*. The pattern of interconnection, that is, the bonding pattern of the density domains is revealed within this range, and, consequently, this density range is also called the *bonding range for density domains*.

If the density threshold is further lowered, then all nuclei of the molecule are found within a common density domain, and the essential molecular pattern of bonding is established; this range is called the *molecular density range*. This molecular density range also contains subranges. At the high threshold values within the molecular range, the density domains usually have some local "neck" regions: there is at least one topological belt or some other multiply connected set on the surface of the density domain along which the density domain $DD(K, a)$ is not locally convex. Within this density subrange the molecular body $DD(K, a)$ is "skinny", and this range is referred to as the *skinny molecular range*. At lower densities no such neck regions occur, but one often finds at least one local nonconvex region along the surface of the density domain $DD(K, a)$. If this occurs within a subrange, then this subrange is called the *corpulent molecular range*. At very low densities the density domains of all molecules (even those of elongated macromolecules) are convex. The corresponding subrange is called the *quasi-spherical molecular range*.

The atomic and the functional group ranges together form the *localized range*, and the molecular density range is regarded as the *global density range*.

The general trends of density domains justify their use as a practical guide for choosing density fragments for building electronic density representations for large molecules. By the application of a simple electron density fragment additivity principle, described earlier [12], it is possible to build rather accurate electron densities of large molecules from molecular fragment densities obtained from accurate calculations for small molecules. These fragment densities can be stored in a database, and used by the MEDLA (Molecular Electron Density "Lego" Assembler) method [12] to "build" electronic densities for various molecules. When selecting molecular fragments as possible building blocks, it is natural to select fragments which possess a separate density domain $DD(K, a)$ for some range of density thresholds a. Hence, density domains can serve as tools for the construction of density databases for later use in the MEDLA method [12], where the "lego"-type fragment combination process has motivated the naming of the technique.

The MEDLA method is based on the following electron density fragment additivity principle. A natural scheme for the implementation of this principle has been described in [12]. Consider an LCAO ab initio wavefunction of a small molecule of some fixed conformation K. If n is the number of atomic orbitals $\varphi_i(r)$ $(i = 1, 2, \ldots, n)$, r is the three-dimensional position vector variable, and P is the $n \times n$ density matrix, then the electronic density $\rho(r)$ of the molecule is given by

$$\rho(r) = \sum_{i=1}^{n} \sum_{j=1}^{n} P_{ij}\, \varphi_i(r)\, \varphi_j(r). \tag{5}$$

An arbitrary collection of the nuclei from the molecule, and the elements of the above density matrix are used to define an electronic density fragment of the molecule. In order to define k-th fragment $\rho^k(r)$ of the electron density $\rho(r)$, the following criterion is used to generate the $n \times n$ fragment density matrix P_{ij}^k:

$P_{ij}^k = P_{ij}$ if both $\varphi_i(r)$ and $\varphi_j(r)$ are AOs centered on nuclei of the fragment,

$\quad = 0.5\, P_{ij}$ if precisely one of $\varphi_i(r)$ and $\varphi_j(r)$ is centered on a nucleus of the fragment,

$\quad = 0$ otherwise. $\tag{6}$

Note that both the complete density matrix $\boldsymbol{P}$ and the fragment density matrix $\boldsymbol{P}^k$ have the same dimensions, $n \times n$.

The electron density of the k-th fragment is defined by the fragment density matrix $\boldsymbol{P}^k$, as

$$\rho^k(\boldsymbol{r}) = \sum_{i=1}^{n} \sum_{j=1}^{n} P_{ij}^k \varphi_i(\boldsymbol{r}) \varphi_j(\boldsymbol{r}). \tag{7}$$

Divide the nuclei of the molecule into m mutually exclusive groups, in order to generate m fragments. The sum of the fragment density matrices is equal to the density matrix of the molecule, and the sum of the fragment densities is equal to the density of the molecule:

$$P_{ij} = \sum_{k=1}^{m} P_{ij}^k \tag{8}$$

and

$$\rho(\boldsymbol{r}) = \sum_{k=1}^{m} \rho^k(\boldsymbol{r}). \tag{9}$$

The above simple additivity rules of Eqs. (8) and (9) are exact on the given ab initio LCAO level; that is, the rebuilding of the electronic density of the given small molecule is exact.

One advantage of the above schemes lies in the possibility of building approximate electron densities for large molecules. Such fragments can be combined to form approximate electron density for a different molecule by selecting and arranging fragment densities so that the nuclear positions closely match those in the target molecule. This procedure, called the Molecular Electron Density "Lego" Assembler (MEDLA) approach, has been implemented in a computer program, MEDLA 93 [15], and has been shown [12] to produce approximate electron densities that are quantitatively very similar to densities obtained in direct 6-31G** ab initio calculations. The molecular isodensity contours (MIDCOs) obtained by the MEDLA and direct ab initio methods are found visually indistinguishable for a family of small molecules. The MEDLA method extends the scope of earlier local and global shape analysis methods of molecules [4] to macromolecules, such as large biomolecules, peptides and proteins.

The scheme of Eq. (6) is not the only one that can be used within the additive framework of Eqs. (7)–(9). The interfragment electron density represented by the interfragment density matrix elements weighted by 0.5 in the scheme of Eq. (6) can be distributed by other weighting schemes. For example, a scheme may be based on comparisons of fragment charges calculated in the parent molecules, or on simple electronegativity comparisons. In general, the following scheme covers most alternatives:

$P^k_{ij} = P_{ij}$ if both $\varphi_i(r)$ and $\varphi_j(r)$ are AOs centered on nuclei of the fragment,

$= f(k,i,j) P_{ij}$ if precisely one of $\varphi_i(r)$ and $\varphi_j(r)$ is centered on a nucleus of fragment k, where $f(k,i,j) > 0, f(k,i,j) + f(k',i,j) = 1$, and where the fragment k' contains the other nucleus,

$= 0$ otherwise. (10)

Take any scalar property $A(i)$ that can be assigned to atomic orbitals and does not change sign, for example, appropriately rescaled electronegativity. For each such scalar property $A(i)$, a simple function $f(k,i,j)$ can be chosen as

$$f(k,i,j) = A(i)/[A(i) + A(j)], \tag{11}$$

where we have assumed that orbital $\varphi_i(r)$ is centered on a nucleus that belongs to the fragment k.

In any of these alternatives, it is expected that those fragments serve best within the MEDLA approach which possess a well recognizable identity, as manifested by the existence of a single density domain for the nuclei of the fragment within the parent molecule.

2 Topological Sequences of Molecular Range Density Domains

The direct, topological characterization of the sequence of density domains $DD(K,a)$ as the density threshold a is gradually decreased has been suggested earlier [4, 8]. For small molecules, the bonding pattern is usually simple, and such a sequence contains only few topologically different members. Consequently, this approach could provide only a rough tool for similarity analysis of small molecules: if the sequences for two molecules are topologically equivalent, then the two molecules are similar according to this rough criterion. For small molecules, a detailed curvature analysis, followed by a topological character-ization of the curvature patterns within the Shape Group Method (SGM) provided a more useful basis for similarity analysis. However, with the introduc-tion of the MEDLA approach for large molecules such as proteins [12], the direct, topological analysis of density domains [4, 8] has acquired new signifi-cance. Clearly, a detailed curvature analysis of all density contours of an entire protein is much too complex and also unnecessary for most purposes, whereas the topological changes of density domains already provide enough variety and complexity for molecular similarity analysis.

Whereas the topological changes throughout the whole range of density thresholds provide more detail, here we shall be concerned only with the changes within the molecular density range. This restriction has some practical advan-tages: within the molecular range there is a single molecular body and there is no need to consider disconnected parts.

When gradually reducing the density threshold within the molecular range, the single density domain $DD(K,a)$ undergoes a series of topological changes, represented by the finite sequence of topologically different density domains

$$DD_1(K,a_1), DD_2(K,a_2), DD_3(K,a_3), \ldots, DD_z(K,a_z). \tag{12}$$

Each of these density domains $DD_i(K,a)$ can be characterized by its three homology groups, H^0, H^1, and H^2. Characterizations by the one-dimensional homology group H^1 (or, by an alternative but similar approach, by the fundamental group, the one-dimensional homotopy group) are the most important:

$$H_1^1(K,a_1), H_2^1(K,a_2), H_3^1(K,a_3), \ldots, H_z^1(K,a_z). \tag{13}$$

The one-dimensional Betti numbers

$$B_1(K,a_1), B_2(K,a_2), (B_3(K,a_3), \ldots, B_z(K,a_z), \tag{14}$$

are the ranks of these groups H^1, where the superscript of dimension is supressed in the notation. The sequence of these Betti numbers, considered as a vector, gives a numerical shape code for the sequence of Eq. (12) of density domains:

$$(B_1(K,a_1), B_2(K,a_2), B_3(K,a_3), \ldots, B_z(K,a_z)). \tag{15}$$

A simple comparison of these numerical sequences for two molecules provides a numerical similarity measure. One of the simplest of these measures for two different conformations K and K' of a molecule is the number of element matches divided by the number of elements in the longer of the two vectors:

$$\text{sim}(K,K') = (1/\max(z,z')) \sum_{i=1}^{\min(z,z')} (1 - \text{abs}(\text{sign}(B_i(K,a_i)$$

$$- B_i(K',a_i)))). \tag{16}$$

3 Topological Patterns of Complete Density Domain Sequences

For the analysis of DD patterns and their changes in conformational processes, we shall adapt a technique originally described for patterns of nuclear potentials of molecules [16]. Consider a molecule containing n' nuclei, and order these nuclei into a sequence according to decreasing nuclear charge:

$$Z_1, Z_2, \ldots, Z_{n'}. \tag{17}$$

For the conformational reaction path p considered, denote the initial nuclear configuration by

$$K_0 = p(0). \tag{18}$$

For an initial choice, take a high enough DD threshold value a_1 that fulfills the following two conditions:

(i) each maximum connected component of the $DD(K_0, a_1)$ set contains precisely one nucleus,
(ii) the $DD(K_0, a_1)$ set has the maximum number of maximum connected components, subject to condition (i).

For such a threshold, the various nuclear neighborhoods are not joined yet (condition (i)), and $DD(K_0, a_1)$ has the maximum number of such atomic neighborhoods (condition (ii)).

We shall denote the n maximum connected components of the corresponding $DD(K_0, a_1)$ by

$$DD_{11}(K_0, a_1), DD_{12}(K_0, a_1), \ldots, DD_{1n}(K_0, a_1). \tag{19}$$

At the threshold a_1, the DD component enclosing nucleus j is denoted by $DD_{1j}(K_0, a_1)$.

All these DD components are topological spheres. This family of DD components is represented by the numerical sequence

$$k_{11}, k_{12}, \ldots, k_{1m}, \tag{20}$$

where k_{1j} is defined by

$$k_{ij} = \begin{cases} Z_j, & \text{if the } j\text{-th nucleus is enclosed by a component of } DD(K_0, a_1) \\ 0 & \text{otherwise} \end{cases}$$

$$\tag{21}$$

As the DD threshold a_1 is decreased to a new value a_2,

$$a_1 > a_2, \tag{22}$$

some of the components $DD_{1j}(K_0, a_1)$ and $DD_{2j}(K_0, a_1)$ may expand and join each other to form a single maximum connected $DD_{2j}(K_0, a_2)$. Here we assume that

$$j < j'. \tag{23}$$

If there are more than two components joining simultaneously, for example, due to symmetry, then the index j of the new, single maximum connected component $DD_{2j}(K_0, a_2)$ is the smallest value of the indices of the joining components. We choose the value a_2 as the DD threshold where the first such joining of components occurs.

In general, a new threshold value a_i, called a *critical threshold value*, is specified for each occasion where a topologically significant change of the DD components occurs, that is, where two or more components join or where their topology changes, for example, where a toroidal DD becomes a topological sphere. These critical threshold values form a finite sequence,

$$a_1, a_2, \ldots, a_m, \tag{24}$$

where the number

$$m = m(K) = m(K_0) \tag{25}$$

is equal to the total number of topological types of the DDs occurring at nuclear configuration $K = K_0$. Some topological types may occur repeatedly along the sequence, as long as some other type occurs for some intermediate threshold value.

For each critical DD threshold value a_i, and for the corresponding family of maximum connected components $DD_{ij}(K_0, a_i)$ of $DD(K_0, a_i)$, a separate more general numerical sequence is specified. For the numbers of the sequence

$$k_{i1}, k_{i2}, \ldots, k_{in}, \tag{26}$$

the relation

$$\mathrm{Re}(k_{ij}) = Z_{j'}, \tag{27}$$

holds, where index j' is the smallest index of any nucleus enclosed by the component $DD_{ij}(K_0, a_i)$ enclosing nucleus j, and where

$$\mathrm{Im}(k_{ij}) = \mathrm{genus}(DD_{ij}(K_0, a_i)). \tag{28}$$

According to this convention, the imaginary component of each number $k_{11}, k_{12}, \ldots, k_{1n}$ of the first sequence is zero that *agrees* well with the assumption on the topological sphere properties of the initial DD components. The numbers $k_{11}, k_{12}, \ldots, k_{1n}$ are real integers which satisfy the conditions of the general definition given for the numerical sequence $k_{i1}, k_{i2}, \ldots, k_{in}$ of a generic index i. In a general sequence of index i, the numbers $k_{i1}, k_{i2}, \ldots, k_{in}$ are complex with integer components, possibly but not necessarily of zero imaginary parts. For indices satisfying Eq. (27), the relation

$$DD_{ij'}(K_0, a_i) = DD_{ij}(K_0, a_i) \tag{29}$$

holds.

3.1 Matrix Representations of Density Domain Sequences

For each nuclear configuration K there are $m(K)$ such $k_{i1}, k_{i2}, \ldots, k_{in}$, sequences, $i = 1, 2, \ldots m(K)$. These sequences can be arranged into matrices of dimensions $m(K) \times n$, which can be augmented by $m_{\max} - m(K)$ rows of additional zeroes, where

$$m_{\max} = \max\{m(K), K \in P\}. \tag{30}$$

Having a common dimension for all the relevant matrices along the conformational path p simplifies the analysis. The resulting, augmented $m_{\max} \times n$

matrices $\mathbb{D}(K)$ have the general form

$$\mathbb{D}(K) = \begin{matrix} k_{11} & k_{12} & \cdots & k_{1n} \\ \cdot\cdot & \cdot\cdot & \cdots & \cdot \\ k_{i1} & k_{i2} & \cdots & k_{in} \\ \cdot\cdot & \cdot\cdot & \cdots & \cdot \\ k_{m(K)1} & k_{m(K)2} & \cdots & k_{m(K)n} \\ 0 & 0 & \cdots & 0 \\ \cdot\cdot & \cdot\cdot & \cdots & \cdot \\ 0 & 0 & \cdots & 0 \end{matrix} \tag{31}$$

Considering a single, static nuclear configuration K, the matrix $\mathbb{D}(K)$ describes the pattern of the topological structure of the density domains.

In order to augment the information contained in matrix $\mathbb{D}(K)$, an m_{max}-dimensional vector $a(K)$ containing the critical threshold values $a, a_2, \ldots, a_m$, as well as an appropriate number of zero entries as elements can also be specified,

$$a(K) = (a_1, a_2, \ldots, a_m, 0, \ldots 0)'. \tag{32}$$

In general, we consider column vectors and the symbol $'$ stands for transpose.

The pair composed of matrix $\mathbb{D}(K)$ and vector $a(K)$ provides a more detailed description of the shape of the DD sequence of nuclear configuration K.

The shape information stored in matrix $\mathbb{D}(K)$ can also be represented by a labeled graph $d(K)$. There are n vertices of graph $d(K)$, one for each nucleus. These vertices are labeled by the serial indices of the nuclei (the column index j' of matrix matrix $\mathbb{D}(K)$). Furthermore, each vertex j' is labeled by a sequence of complex numbers $z_{j't}$, $t = 1, 2, \ldots$, with real parts

$$\text{Re}(z_{j't}) = i', \tag{33}$$

where i' is the index of density threshold value $a_{i'}$ where the t-th topological change of $DD_{i'j'}(K_0, a_{i'})$ occurs, and with imaginary parts

$$\text{Im}(z_{j't}) = \Delta(\text{genus}), \tag{34}$$

representing the topological change, as long as for this change j' is the smallest nucleus index within the density domain DD.

The edges of graph $d(K)$ are defined by the following condition. There is an edge from vertex j to vertex j' if at the nuclear potential threshold a_i the nucleus j is contained in the DD component $DD_{ij'}(K_0, a_i)$, where j' is the smallest index of any nucleus enclosed by the DD component containing nucleus j, and where a_i is the largest threshold value where this holds. For simplicity, the (j, j') edge is labeled by index i.

The (j, j') edge indices can also be used to assign a direction to each edge, for example, the direction from higher to lower nuclear index as given in the list at Eq. (20), turning the edges into arcs and the graph $d(K)$ into a digraph. The digraph $d(K)$ is a discrete representation of the topological pattern of the density domains of the molecule.

Another digraph representation $d_a(K)$ can be obtained from digraph $d(K)$ if the integer arc labels i are replaced with the real number label a_i and the real parts i' of the arc labels are replaced with the actual threshold values $a_{i'}$, where the topological changes occur. This alternative approach takes into account all information represented by matrix $\mathbb{D}(K)$ and vector $a(K)$. Of course, digraph $d_a(K)$ is no longer a discrete representation of the density domains. For simplicity in the following discussion we shall use the matrix representations $\mathbb{D}(K)$ and vectors $a(K)$, convenient for computer manipulations.

3.2 Topologically Significant Shape Changes of DD Sequences along Reaction Paths

In chemical reactions and conformational changes, such as the process modeled by the formal reaction path p, the matrix $\mathbb{D}(K)$ as well as the vector $a(K)$ change as the nuclear configuration K varies. Consider the path p as a formal mapping from the unit interval I to the metric nuclear configuration space M [13],

$$p: I = [0, 1] \rightarrow M. \tag{35}$$

This mapping represents a parametrization,

$$p = p(u), \quad 0 \leq u \leq 1, \tag{36}$$

where $u = 0$ corresponds to the initial point

$$K_0 = p(0) \in M, \tag{37}$$

and the choice $u = 1$ corresponds to the "final" nuclear configuration

$$K_f = p(1) \in M \tag{38}$$

of the formal product.

A small displacement Δu does not necessarily alter the topological pattern $\mathbb{D}(K(u))$ of DD sequences for most nuclear configurations $K(u)$ along the path $p(u)$; in fact, for most small displacement Δu,

$$\mathbb{D}(K(u)) = \mathbb{D}(K(u + \Delta u)). \tag{39}$$

Note, however, that the numerical values of the critical thresholds that are usually non-integer real numbers stored in vector $a(K)$, are likely to change:

$$a(K(u)) \neq a(K(u + \Delta u)). \tag{40}$$

As a consequence of the discrete matrix representation, along the entire path $p(u)$, there are only a finite number of different $\mathbb{D}(K(u))$ matrices of DD sequences.

Consequently, the path $p(u)$ can be decomposed into a finite number w of invariance intervals,

$$p_{D,1}, p_{D,2}, \ldots, p_{D,w}. \tag{41}$$

The corresponding sequence

$$\mathbb{D}(p,1), \mathbb{D}(p,2),\ldots, \mathbb{D}(p,w) \tag{42}$$

of matrices, in combination with the $w - 1$ parameter values

$$u_{D,1}, u_{D,2},\ldots u_{D,w-1} \tag{43}$$

marking the endpoints of the first $w - 1$ of the invariance intervals $p_{D,1}$, $p_{D,2},\ldots p_{D,w-1}$ along path p, can be used to characterize the topological pattern of density domains and its variation in the course of the chemical process represented by p.

The parametrization of a given physical path is not unique. However, it is always possible to reparametrize the path p in such a way that the actual $u_{D,1}, u_{D,2},\ldots u_{D,w-1}$ values change while preserving their monotonic increase along the path p. In particular, if the parameterization of Eq. (36) reflects some physical condition, for example, if it is based on proportionality with the distance between configuration as given in the metric nuclear configuration space M [13], then the $u_{D,1}, u_{D,2},\ldots u_{D,w-1}$ parameter values also reflect information of direct physical significance.

3.3 Density Domain Shape Similarity of Reaction Paths

The above tools are also suitable for studying the similarity of reaction paths. Two reaction paths p_1 and p_2 are regarded shapewise equivalent within the above context ($\mathbb{D}$-shape equivalent) if and only if the numbers w_1 and w_2 of their shape invariance intervals agree,

$$w_1 = w_2 = w, \tag{44}$$

and if the two matrix sequences

$$\mathbb{D}(p_1,1), \mathbb{D}(p_1,2),\ldots, \mathbb{D}(p_1,w) \tag{45}$$

and

$$\mathbb{D}(p_2,1), \mathbb{D}(p_2,2),\ldots, \mathbb{D}(p_2,w) \tag{46}$$

are the same,

$$\{\mathbb{D}(p_1,1), \mathbb{D}(p_1,2),\ldots, \mathbb{D}(p_1,w)\} = \{\mathbb{D}(p_2,1), \mathbb{D}(p_2,2),\ldots, \mathbb{D}(p_2,w)\}. \tag{47}$$

The above shapewise equivalence of two reaction path p_1 and p_2 according to the matrix sequence of Eq. (45) is denoted by

$$p_1 \, \mathbb{D} \, p_2. \tag{48}$$

This equivalence relation generates equivalence classes denoted by P_1, where the index 1 is inherited from the arbitrarily chosen representative path p_1 of the class P_1. Of course, for the above two paths

$$p_1, p_2 \in P_1. \tag{49}$$

The shape-based similarities between various reaction mechanisms can be analysed and quantified by direct comparisons of their matrix sequences. A numerical similarity measure of reaction mechanisms is based on a measure of difference between the two matrix sequences: the smaller the difference, the greater the similarity. The extreme case of shapewise equivalence of reaction mechanisms is represented by Eq. (47).

One can use a derivation entirely analogous with the homotopy equivalence classes of paths and loops and the fundamental group of reaction mechanisms in the nuclear configuration space M [13], leading to a group theoretical model of reaction mechanisms based on shape. The above shape equivalence of reaction paths p in space M generates a complete shape classification of all possible reaction paths for the given stoichiometry of nuclei.

Another alternative is provided by considerations of shape variations within conformational domains instead of along reaction paths. Formal chemical species are represented by various catchment regions $C(\lambda, i)$ of the nuclear configuration space M, with respect to a specified electronic state and the associated potential energy hypersurface $E(K)$ [13]. A catchment region is defined as the collection of all nuclear configurations that are starting points of steepest descent paths leading to a common critical point of the given potential surface. For a catchment region $C(\lambda, i)$, λ is the index of the critical point (in particular, $\lambda = 0$ for a minimum and $\lambda = 1$ for a simple saddle point of a transition structure), whereas i is simply a serial index.

Take a subclass $P_1(C(\lambda, i), C(\lambda', i'))$ of class P_1, defined by the following condition: $P_1(C(\lambda, i), C(\lambda', i'))$ is the family of all paths from class P_1 which start at the catchment region $C(\lambda, i)$, end at the catchment region $C(\lambda', i')$, and are homotopic to one another (continuously deformable into one another) while preserving these properties. Evidently the above conditions correspond to an equivalence relation among paths, and, consequently, $P_1(C(\lambda, i), C(\lambda', i'))$ is an equivalence class. Such equivalence classes $P_1(C(\lambda, i), C(\lambda', i'))$ represent formal *reaction mechanisms defined in terms of the shape of Density Domains* (D-shape).

Using representative reaction paths for reaction mechanisms, the similarities between various reaction mechanisms can be analysed and quantified by direct comparison of their matrix sequences $\mathbb{D}(p, 1), \mathbb{D}(p, 2), \ldots, \mathbb{D}(p, w)$. A numerical similarity measure of reaction mechanisms is given by a measure of difference between the two matrix sequences. A smaller difference between the two matrix sequences implies a greater similarity of the reaction mechanisms. Equation (47) represents the extreme case of similarity: the shapewise equivalence of reaction mechanisms.

3.4 Topological Varieties of Shapes of Density Domains Within Conformational Ranges; DD Shape Invariance Domains of the Configuration Space

If one studies the range of molecular deformations which preserve chemical identity, or the range of molecular motions preserving some other physical property, it is often inconvenient to rely on formal conformational paths or reaction paths in configuration space M. The preserved physical property itself can be used to specify various domains of the configuration space M, and the shape features occurring within these domains can be studied without reference to formal paths. The catchment regions $C(\lambda, i)$ of the configuration space M with respect to a specified electronic state and the associated potential energy hypersurface $E(K)$ have been suggested as natural representatives of configurational families preserving chemical identity [13]; the boundaries of these catchment regions mark the limits of deformations of the given chemical species.

The nuclear configuration space M can be subdivided into topological shape invariance domains as specified by the shape groups of the electronic density. A somewhat more general representation is obtained if the configuration space M is combined with the continuous parameters of the shape characterization technique, such as the density threshold a and the reference curvature parameter b of the shape group method, generating the dynamic shape space D used as a tool for dynamic shape analysis [17]. The dynamic shape space approach leads to a detailed shape description for the specified electronic state. A similar approach can be adapted to the topological patterns of density domains, focussing on the matrix representations described above.

Using Eq. (31) with a modified choice for m_{max},

$$m_{max} = \max\{m(K), K \in M\}, \tag{50}$$

a new DD matrix $\mathbb{D}(K)$ can be defined for each nuclear configuration K. These matrices are invariant for most small variations of the nuclear configuration K. Except for some pathological cases, there are only a finite number of $\mathbb{D}(K)$ matrices associated with any specified electronic state and the nuclear configuration space M. The invariance domains of these matrices are the density domain shape invariance domains of the metric nuclear configuration space M.

The finite number q of different matrices $\mathbb{D}(K)$ can be listed according to some representative nuclear configurations

$$K_1, K_2, \ldots, K_q, \tag{51}$$

and the associated list of the finite number of DD shape invariance domains by the above matrix criterion is given as

$$M_{D,1}, M_{D,2}, \ldots, M_{D,q}. \tag{52}$$

These domains represent a partitioning of M, hence the union of these DD shape invariance domains is the entire nuclear configuration space M,

$$M = \cup M_{D,k}. \tag{53}$$

The matrices $\mathbb{D}(K_k)$ can be used to specify *shape types*: we say that a nuclear configuration is of shape type k ($\mathbb{D}$-shape type k), if K belongs to the equivalence class $M_{D,k}$.

The nuclear configuration space M is provided with a metric that allows one to introduce a measure of volume V for various subsets of M. The relative importance of a given property can be characterized by the volume of the subset of M where this property is present. The relative importance of a given shape as specified by the associated matrix $\mathbb{D}(K_k)$ can be expressed by volumes $V(M_{D,k})$ of invariance domains $M_{D,k}$.

In particular, the contribution of a specified shape type to a given chemical species is of relevance: a chemical species that exhibits a given shape type for most of its distorted configurations is a more reliable representative of this particular shape than another molecule that has only a much smaller set of distorted forms with this shape type. In fact, a measure of the relative importance of a specified shape type k ($\mathbb{D}$-shape type k) for an individual chemical species $C(\lambda,i)$ can be expressed by the number $s_c(k,i)$, defined as the following volume ratio:

$$s_c(k,i) = V(M_{D,k} \cap C(\lambda,i))/V(C(\lambda,i)). \tag{54}$$

Similarly, the contribution $c_s(i,k)$ of chemical species $C(\lambda,i)$ to a given shape type k ($\mathbb{D}$-shape type k) is expressed as the following volume ratio:

$$c_s(i,k) = V(M_{D,k} \cap C(\lambda,i))/V(M_{D,k}). \tag{55}$$

According to Eq. (53), the shape invariance domains $M_{D,k}$ give a complete partitioning of the configuration space M. Consequently these invariance domains restricted to any catchment region $C(\lambda,i)$ must also give a complete partitioning of the catchment region. As a result, the relation

$$\sum_k s_c(k,i) = 1, \tag{56}$$

must hold.

In all but some pathological cases, the catchment regions give a complete partitioning of the configuration space M (see [13] for some exceptions). This implies that the catchment regions $C(\lambda,i)$, restricted to any invariance domain $M_{D,k}$, must also give a complete partitioning of the invariance domain $M_{D,k}$. As a consequence, one obtains

$$\sum_i c_s(i,k) = 1, \tag{57}$$

where the index i in $c_s(i,k)$ carries the implicit λ-dependence.

The two relations of Eqs. (56) and (57) can be used as a test for results obtained for individual shape and species contributions.

4 Local Shape Invariance of Density Domains and the Transfer of Functional Groups in Reactions

There are many regularities and similar features in the chemical properties of the more common functional groups in chemistry. In earlier studies [18] a general framework has been developed for the analysis of the interrelations and transformations of flexible functional groups within the configuration space M of all possible chemical species of fixed overall stoichiometry S. This general approach is applicable for functional groups defined in terms of density domains.

The flexibility of various functional groups can be characterized by topological criteria, and can be compared to the fundamental pattern of the distribution of configurational families within the nuclear configuration space M. This pattern gives a concise description of the occurrence and interconversion of various functional groups in families of nuclear arrangements during chemical reactions and conformational changes, where the identity of a molecules and their functional groups is specified in terms of the given topological criteria.

Depending on the actual distortion and strains in a molecule, the identity of a functional group may or may not be preserved. A large enough local distortion in a molecular moiety may qualify as a formal chemical reaction that changes a functional group into another. For example, a large enough local distortion may convert an enol group into a keto group, or to take a more extreme example, the cis-CH$=$CFH group into the trans-CH$=$CFH group.

In [18] a rather general framework has been proposed for such problems, and below we shall outline the main features of this approach, with special emphasis on applications to density domains.

Consider a stoichiometry S and the nuclear configuration space M associated with this stoichiometry. The DD functional groups can be modeled by the MEDLA MIDCO approach [12], and their shapes can be characterized by the matrix technique described earlier. Take some family f of DD functional groups f_t generated by some or all of the N nuclei specified for the given stoichiometry S:

$$f = \{f_1, f_2, \ldots, f_t, \ldots, f_w\}. \tag{58}$$

We shall use the term "functional group" in a very general sense; among these w functional groups in set f we may include two extreme types of chemical entities, such as individual atoms, and entire molecules, possibly containing all of the N atoms of the stoichiometry S. This rather broad interpretation of the term "functional group" is convenient when formulating general rules.

A functional group involving many nuclei may contain some smaller functional groups of fewer nuclei. For example, the

$$f_s = -CH_2 - CH_2 - NH_2 \tag{59}$$

group contains the

$$f_t = -CH_2 - NH_2 \tag{60}$$

group. This fact can be regarded as a *chemical inclusion relation* that we formally denote by

$$f_t < f_s. \tag{61}$$

The chemical inclusion relation is not necessarily applicable for an arbitrary pair of DD functional groups. In fact, for most pairs of DD functional groups of such a family f no such chemical inclusion relation exists. For example, this is the case for the functional groups $f_t' = -CH_2 - NH_2$ and $f_s' = -CH_2 - CH_3$: neither DD functional group contains the other. Consequently, the chemical inclusion relation > defines only a *partial order* within the family f.

In some important cases, two DD functional groups are of special interest. For many choices of set f, there exists a functional group f_1 which is contained in all the other functional groups $f_2, \ldots, f_w$ of the family f,

$$f_1 < f_s, \quad f_s = f_2, \ldots, f_w. \tag{62}$$

If the relations of Eq. (62) hold, then this functional group f_1 is called an *infimum* with respect to the chemical inclusion relation < taken as the partial order. In this case the family f is a *lower semilattice*. For example, the family f can be chosen so that each member f_s contains a common atom, for example, the H atom, where H is also regarded as a functional group. Then the < partial order relation of chemical inclusion implies that family f is a lower semilattice, with $f_1 = H$ as its infimum element [18].

Another type of functional group of special importance is one that contains many other functional groups of the family f. The extreme case corresponds to another algebraic elements with respect to the chemical inclusion relation. If the set f contains only one, unique chemical species represented by a DD that involves all the N nuclei of the given stoichiometry S, where this species itself is regarded as a formal DD functional group f_w, and if f_w contains all other functional groups of the family f,

$$f_j < f_w, \quad f_s = f_1, \ldots, f_{w-1}, \tag{63}$$

then the functional group f_w can be taken as a *supremum* with respect to the chemical inclusion relation < taken as the partial order. In this case, family f of the DD functional groups is an *upper semilattice*.

In special cases, both relations of Eqs. (62) and (63) hold, that is, both infimum and supremum exists within family f with respect to the < partial order relation. In such a case the family f of functional groups is a *lattice*. Lattices and semilattices are important algebraic tools for systematic analyses of various hierarchies.

The general scheme of the earlier study [18] gave a tool for the analysis of conditions for the presence and interrelations of functional groups among all possible nuclear configurations of a given stoichiometry S. Recognizing the nonrigid, flexible nature of functional groups, a topological "tolerance" range has been selected for these groups in order to account for the allowed, minor geometry changes which do not change their chemical identity. According to these criteria, if two geometrical arrangements of a given collection of atoms are similar enough, then these two arrangements are considered as representations of the *same* functional group. The criteria may vary for the functional groups: for each functional group f_t a range T_t of allowed geometrical arrangements is specified and within the given range each functional group f_t is regarded to preserve its chemical identity. In the earlier study [18], the family of all these ranges T_t has been denoted by T,

$$T = \{T_1, T_2, \ldots, T_t, \ldots, T_w\}. \tag{64}$$

Each range T_t can be regarded as a family of nuclear configurations, and taken as a topological "tolerance specification" of the geometrical distortions preserving the chemical identity of functional group f_t. If these families of nuclear configurations are mutually disjoint, then all geometrical realizations of a functional group f_t with geometrical arrangements falling within the limitations specified by set T_t can be regarded as topologically equivalent. In the example of the earlier study [18], the tolerance criterion for distortions was selected as any nuclear displacement equal to 25% of the van der Waals atomic radius for each atom.

In terms of density domains of functional groups, alternative sets of topological criteria can be selected, reflecting the preservation of some essential properties of functional groups. These criteria can be specified in terms of the matrix conditions described in earlier parts of this chapter. Here one may exploit the advantages of the MEDLA representations of density domains. In a density domain analysis, molecular fragments can be easily identified simply by taking a subset of the nuclei and the density domains generated by this subset [4, 8]. This density domain analysis described for molecular $DD(K, a_i)$ components is directly applicable for functional groups.

5 Concluding Remarks

Following a brief review of the background and basic topological concepts of molecular isodensity contours (MIDCOs) and density domains (DDs), several, related aspects of their roles in similarity analysis are discussed. The special family of molecular range density domains provides a shape characterization of particular importance in macromolecules, where there is often no need for high resolution details of shapes described by curvature based methods such as the Shape Group Method, SGM. By contrast, the topological patterns of complete density domain sequences are suitable for detailed analysis of various parts of a molecule, and are useful in local similarity analyses. Matrix representations of density domain sequences are introduced and the topologically significant shape changes of density domain sequences along reaction paths are described in terms of matrix sequences. These matrices are suitable for the evaluation of density domain shape similarity measures for reaction paths. Methods are described for the study of the topological varieties of shapes of density domains within conformational ranges and DD shape invariance domains of the configuration space are studied. By combining these approaches within a general framework proposed earlier, techniques are described for the analysis of the local shape invariance of density domains representing functional groups and the transfer of these groups in chemical reactions and conformational rearrangements.

The *ab initio* quality MEDLA electron densities of proteins [19] extend the scope of density domain analysis.

Acknowledgment. The original research leading to the developments described in this chapter was supported by both strategic and operating research grants from the Natural Sciences and Engineering Research Council of Canada.

6 References

1. Coppens P, Hall MB (eds) (1982) Electron distribution and the chemical bond. Plenum, New York and London
2. Fliszár S (1983) Charge distributions and chemical effects. Springer, New York
3. Bader RFW (1992) Atoms in molecules: a quantum theory. Oxford University Press, Oxford
4. Mezey PG (1993) Shape in chemistry: An introduction to molecular shape and topology. VCH, New York
5. Richards WG (1983) Quantum pharmacology. Butterworths, London
6. Dean PM (1987) Molecular foundations of drug-receptor interaction. Cambridge University Press, New York
7. Johnson MA, Maggiora GM (eds) (1990) Concepts and applications of molecular similarity. Wiley, New York
8. Mezey PG (1992) J Chem Inf Comp Sci 32: 650
9. Mezey PG (1994) Can J Chem 72: 928
10. Frisch MJ, Head-Gordon M, Trucks GW, Foresman JB, Schlegel HB, Raghavachari K, Robb MA, Binkley JS, González C, DeFries DJ, Fox DJ, Whiteside RA, Seeger R, Melius CF, Baker J, Martin R, Kahn LR, Stewart JJP, Topiol S, Pople JA (1990) GAUSSIAN 90, Gaussian Inc., Pittsburgh, PA
11. Kollman PA (1978) J Am Chem Soc 100: 2974
12. Walker PD, Mezey PG (1993) J Am Chem Soc 115: 12423
13. Mezey PG (1987) Potential energy hyperfsurfaces. Elsevier, Amsterdam
14. Walker PD, Arteca GA, Mezey PG (1990) Program GSHAPE 90, Mathematical Chemistry Research Unit, University of Saskatchewan, Saskatoon, Canada
15. Walker PD, Mezey PG (1990) Program MEDLA 93, Mathematical Chemistry Research Unit, University of Saskatchewan, Saskatoon, Canada
16. Mezey PG (1994) Discrete representations of three-dimensional molecular bodies and their shape changes in chemical reactions. In: Bonchev D (1994) Graph theoretical approaches to chemical reactivity. Kluwer, Dordrecht
17. Mezey PG (1988) J Math Chem 2: 299
18. Dubois J-E, Mezey PG (1992) Int J Quantum Chem 43: 647
19. Walker PD, Mezey PG (1994) Can J Chem 72: 2531

Momentum-Space Electron Densities and Quantum Molecular Similarity

Neil L. Allan[1] and David L Cooper[2]

[1] School of Chemistry, University of Bristol, Cantocks Close, Bristol BS8 1TS, U. K.
[2] Department of Chemistry, University of Liverpool, P.O. Box 147, Liverpool L69 3BX, U.K.

Table of Contents

This Chapter reviews a recent advance in the quantitative estimation of quantum molecular similarity. In this new approach, molecular similarity and dissimilarity indices are obtained from numerical comparisons of *momentum-space* electron densities. Many of the problems associated with

Topics in Current Chemistry, Vol. 173
© Springer-Verlag Berlin Heidelberg 1995

more conventional position-space procedures are avoided and particular emphasis is placed on the variation of the long-range position-space electron density. The momentum-space approach is particularly suited to problems for which the molecular activity depends less on the details of the bonding topology than on features of the long-range slowly-varying valence electron density.

Momentum-space concepts are not, in general, familiar to the chemist and so we outline first the calculation of momentum-space electron densities, $\rho(p)$, from ab initio wavefunctions. The form of $\rho(p)$ for different molecules is discussed, using as examples (i) the ground state of H_2, (ii) bond formation in BH^+, and (iii) the π-orbitals in large conjugated polyenes.

The construction and the evaluation of similarity and dissimilarity indices based on $\rho(p)$ are described in some detail. Examples are presented involving the comparison of (i) the total or total valence electron densities of two molecules, (ii) the densities associated with particular molecular fragments or localised molecular orbitals, and (iii) the densities of two molecular orbitals in the same molecule. Results are reported for the model series (a) $CH_3CH_2CH_3$, CH_3OCH_3 and CH_3SCH_3, and (b) C–H and C–F bonds in hydrofluoromethanes. Finally, two studies involving larger systems are presented. In the first, momentum-space similarity indices are used to rationalise anti-HIV1 virology data for a group of phospholipids. The technique proves to have predictive value for such systems. In the second application, a structure-activity relationship is generated for the hyper-polarisabilities of a range of non-centrosymmetric 1,4-substituted benzene derivatives.

1 Introduction

It has been well-known since the 1920s that electron density can be expressed as a function of the *momenta* of the electrons rather than as a function of their positions. In practice, this representation is most often employed in condensed matter physics and it is encountered much less frequently in chemical problems. Momentum space is very convenient for the interpretation of the Compton and (e, 2e) scattering experiments [1, 2]. Otherwise, despite the pioneering work of Coulson and Duncanson in the 1940s [3], and a few more recent studies [4], little use has been made of momentum space concepts in chemistry.

Discussed here is our novel use of momentum-space electron densities for the quantitative estimation of molecular similarity. Molecular similarity concepts arise most often in biomolecular science, where it is not uncommon for the processes to be extremely complex, partly characterised, or even completely unknown. There is, of course, considerable current interest in the direct evaluation of the free energies of receptor interactions. However, when the active site is not known, for example, one must resort to alternative strategies (structure-activity relationships) which involve a comparison of the structure and properties of active and inactive molecules. A wide range of techniques have been suggested to assess molecular similarity, relating to an equally diverse selection of applications [5–7]. Quantitative approaches include the comparison of position-space (r-space) electron densities [8, 9] or of electrostatic potentials [10], and the topological analysis of the three-dimensional shapes of charge densities [11]. In addition, there has been much work on graph theoretical methods and on database searching [5–7]. Our approach involves a comparison of the *momentum-space* (p-space) electron densities, $\rho(p)$, of the molecules, orbitals or molecular fragments of interest [12–14].

Momentum-space concepts are generally unfamiliar, and so Sect. 2 provides a brief introduction to the calculation and to the form of the p-space wavefunction, $\Psi(p)$, and of the corresponding p-space electron density, $\rho(p)$. We consider some representative molecules, with particular emphasis on the consequences of chemical bonding in momentum space. Section 3 deals with the construction and computation, using ab initio and semi-empirical wavefunctions, of families of similarity and dissimilarity indices. Section 4 describes results for two model series: the first consists of CH_3OCH_3, CH_3SCH_3 and $CH_3CH_2CH_3$, and the second relates to C–F and C–H bonds in hydrofluoromethanes. These studies involve estimating the (dis)similarity of total densities for the molecules, and for molecular fragments, as well as comparisons of localised orbitals in different molecules. Two applications from our recent work are then outlined in Sect. 5. The first of these relates to the similarity of anti-HIV1 phospholipids and is based on the similarity of total densities for fragments of *different* molecules. The second is concerned with hyperpolarisabilities of a series of 1,4-substituted benzene derivatives and involves the similarity of two orbitals in the *same* molecule. Some general remarks concerning our methodology conclude the review.

2 Momentum-Space Electron Densities

2.1 Definitions and Computational Aspects

The momentum-space electron density, $\rho(p)$, is just as straightforward to evaluate as its analogue in position space, $\rho(r)$. The wavefunction in momentum space, $\Psi(p)$, is simply the Fourier transform of the r-space wavefunction $\psi(r)$:

$$\Psi(p) = (2\pi)^{-3/2} \int \psi(r) \exp(-ip \cdot r) \, dr. \tag{1}$$

Direct solution of the Schrödinger equation is possible in momentum-space for simple systems, but it is usually more convenient to start from r-space quantities. We begin here with molecular orbitals (MOs) $\psi_\mu(r)$ formed in the usual way by the overlap of atomic basis functions $\phi_k(r - R_A)$, centred on nuclei A at positions R_A:

$$\psi_\mu(r) = \sum_k c_{\mu k} \phi_k(r - R_A). \tag{2}$$

This is usually the most convenient starting point, because the coefficients $c_{\mu k}$ may be obtained readily from self-consistent field (SCF) calculations on the molecules of interest. In the work described here, ab initio and semi-empirical SCF wavefunctions have been generated using the GAMESS-UK [15] and MOPAC [16] programs, respectively.

Using Eq. (1), the corresponding momentum-space MO wavefunction $\Psi_\mu(p)$ is

$$\Psi_\mu(p) = \sum_k c_{\mu k} \Phi_k(p) \exp(-ip \cdot R_A) \tag{3}$$

where the $\Phi_k(p)$ are the Fourier transforms of the $\phi_k(r)$:

$$\Phi_k(p) = (2\pi)^{-3/2} \int \phi_k(r) \exp(-ip \cdot r) \, dr. \tag{4}$$

Kaijser and Smith [17] have presented analytic forms for many of the $\Phi_k(p)$. Table 1 collects together those used here, corresponding to Slater-type orbitals and to Gaussian-type orbitals in both spherical harmonic and Cartesian form.

The total momentum-space electron density $\rho(p)$ is given by

$$\rho(p) = \sum_{\mu,\nu} D(\mu|\nu) \Psi_\mu(p) \Psi_\nu(p) \tag{5}$$

in which $D(\mu|\nu)$ is the normalised one-particle density matrix. In the special case of SCF molecular orbitals (with occupancy $n_\mu = 0, 1$, or 2), $D(\mu|\nu) = n_\mu \delta_{\mu\nu}$. It is clear from Eq. (5) that $\rho(p)$ is the direct analogue of the r-space electron density $\rho(r)$.

Two simple tests can be used to check the correct evaluation of $\rho(p)$. First, $\rho(p)$, like $\rho(r)$, must be normalised and thus satisfy:

$$\int \rho(p) \, dp = N \tag{6}$$

where N is the total number of electrons. Secondly, the expectation value of the kinetic energy, $\langle T \rangle$, is (in atomic units):

$$\langle T \rangle = \tfrac{1}{2} \int p^2 \rho(p) \, dp \tag{7}$$

where $p = |p|$. The virial theorem states that for any molecule at its equilibrium geometry

$$\langle T \rangle = -E \tag{8}$$

where E is the total energy of the molecule, which we know from the initial r-space SCF calculation.

In passing, we note that Eq. (7) involves the evaluation of a moment of momentum, defined in general as

$$\langle p^m \rangle = \int p^m \rho(p) \, dp \tag{9}$$

in which typical values of m are $-1, 0, 1$ and 2 [18]. For different values of m, each $\langle p^m \rangle$ emphasises a different region of the electron density.

The Fourier transform (Eq. (1)) preserves direction, in the sense that one can refer to components of the total momentum in any particular direction. For example, one can distinguish components along any Cartesian axis, or parallel/perpendicular to a bond or plane. A further consequence of Eq. (1) is that the momentum density $\rho(p)$ possesses the same symmetry elements as its r-space counterpart $\rho(r)$. In addition, in the absence of net translational motion i.e. $\langle p \rangle = 0$, $\psi(r)$ must be real and $\Psi(p)$ and $\Psi(-p)$ must have the same modulus so that

$$\rho(p) = \rho(-p) . \tag{10}$$

Consequently the momentum density of *any* molecule will have inversion symmetry, even if $\rho(r)$ does not. It can be useful to take advantage of this inversion symmetry when integrating functions of $\rho(p)$, such as the generalised overlaps to be described in Sect. 3.

The momentum density $\rho(p)$ falls off very rapidly with p (see Table 1) and attains its largest values at low values of p. Such regions correspond to the slowly-varying outer valence r-space density. In r-space, on the other hand, the form of the electron density is determined largely by the core electrons and hence by the positions of the nuclei, especially the heavier ones (cf. the problem of determining H-atom positions in X-ray diffraction). Even the valence electron density is dominated to a large extent by the nuclear positions. In marked contrast to the position-space density, the momentum density highlights some of the most chemically interesting parts of the electron distribution, without over-emphasising the bonding topology.

We now present three examples to illustrate momentum-space concepts, starting with a simple and familiar system, namely the hydrogen molecule H_2.

Table 1. Analytic forms of normalised p-space atomic basis functions. In each case α is the r-space orbital exponent. Only unique functions are given for the Cartesian Gaussians

	Spherical Slaters			
	$n = 1$	$n = 2$	$n = 3$	$n = 4$
Common factor	$(2\alpha/\pi)^{1/2}(\alpha^2 + p^2)^{-2}$	$(2\alpha/3\pi)^{1/2}(\alpha^2 + p^2)^{-3}$	$(\alpha/5\pi)^{1/2}(\alpha^2 + p^2)^{-4}$	$(2\alpha/35\pi)^{1/2}$ $\times(\alpha^2 + p^2)^{-5}$
$l = 0$	$4\alpha^2 Y_{00}$	$4\alpha^2(3\alpha^2 - p^2)Y_{00}$	$32\alpha^4(\alpha^2 - p^2)Y_{00}$	$16\alpha^4(p^4 - 10p^2\alpha^2 + 5\alpha^4)Y_{00}$
$l = 1$	—	$-16i p\alpha^3 Y_{1m}$	$-(32/3)i p\alpha^3$ $\times(5\alpha^2 - p^2)Y_{1m}$	$-32i p\alpha^5(5\alpha^2 - 3p^2)Y_{1m}$
	Cartesian Gaussians			
Common factor	1s	$2p_x$	$3d_{xx}$	$3d_{xy}$
$(2\alpha\pi)^{3/4}e^{-p^2/4\alpha}$	1	$-ip_x/\sqrt{\alpha}$	$-\dfrac{1}{\sqrt{3}}\left(\dfrac{p_x^2}{\alpha} - 2\right)$	$-\dfrac{p_x p_y}{\alpha}$
	Spherical Gaussians			
Common factor	1s	2p	3d	4f
$(2\alpha/\pi)^{1/4}e^{-p^2/4\alpha}$	$\dfrac{1}{\alpha}Y_{00}$	$-\dfrac{ip}{\alpha\sqrt{3\alpha}}Y_{1m}$	$-\dfrac{p^2}{\sqrt{15\alpha^2}}Y_{2m}$	$\dfrac{ip^3}{\alpha^2\sqrt{105\alpha}}Y_{3m}$

2.2 The Hydrogen Molecule H_2

At the simplest level, the doubly-occupied $1\sigma_g$ molecular orbital in H_2 is formed from the overlap of atomic 1s functions centred on the two nuclei A and B, so that:

$$\psi(r) = \frac{1}{\sqrt{2(1 + S)}} (\phi_{1s}(r - R_A) + \phi_{1s}(r - R_B)) \tag{11}$$

in which S is the overlap integral defined in the usual way. Using Eq. (1),

$$\Psi(p) = \frac{1}{\sqrt{2(1 + S)}} \Phi_{1s}(p)(\exp(-ip \cdot R_A) + \exp(-ip \cdot R_B)) \tag{12}$$

where $\Phi_{1s}(p)$, a real function, is the Fourier transform of $\phi_{1s}(r)$.

The momentum density (Eq. (5)) $\rho(p)$ follows at once

$$\rho(p) = \Psi^*(p)\Psi(p)$$

$$= \frac{\Phi_{1s}^2(p)}{1 + S}(1 + \cos(p \cdot R_{AB})) \tag{13}$$

in which $R_{AB} = R_A - R_B$ is origin-independent. Substituting explicitly for $\Phi_{1s}(p)$ from Table 1,

$$\rho(p) = \frac{32\alpha^5}{(1 + S)\pi(\alpha^5 + p^2)^4}(1 + \cos(p \cdot R_{AB})). \tag{14}$$

This last equation illustrates our earlier remark that the momentum density falls off rapidly with p (in the present case as p^8 at large p). There is a maximum in $\rho(p)$ for this orbital at $p = 0$. This can be seen in Fig. 1b, which shows an isometric view of $\rho(p)$ in the plane $p_x = 0$. Figure 1a is the analogous r-space plot of $\rho(r)$ in the plane $x = 0$. In both cases, the bond direction is the z-axis. In fact, Fig. 1 was generated using a somewhat more elaborate basis set (TZVP) than was assumed in our simple algebraic treatment, but all the features we discuss are common to the densities derived from both calculations.

The term $\cos(p \cdot R_{AB})$ in Eq. (14) introduces oscillatory structure into the momentum density. Using the orientation of the axes defined in Fig. 1,

$$\rho(p) = \frac{2}{(1 + S)} \Phi_{1s}^2(p) \cos^2(\tfrac{1}{2}p_z R_{AB}). \tag{15}$$

Fig. 1a–c. Isometric views of the electron density for H_2. The internuclear axis is the z-axis, with the nuclei at $(0, 0, \pm 0.7)$ bohr: **a** position-space electron density in the plane $x = 0$, extending to ± 1.5 atomic units in both the y and z directions; **b** momentum-space electron density in the plane $p_x = 0$, extending to ± 1 atomic units in both the p_y and p_z directions, with $p = 0$ at the centre of the plot; **c** momentum-space electron density in the plane $p_x = 0$, extending to ± 6 atomic units in both the p_y and p_z directions; the peak around $p = 0$ (at the centre of the plot) has been removed in order to highlight the nodal planes in $\rho(p)$ (Eq. (15)) at $p_z = \pm \pi/R_{AB}$ (i.e. $p_z \approx \pm 2.2$ atomic units)

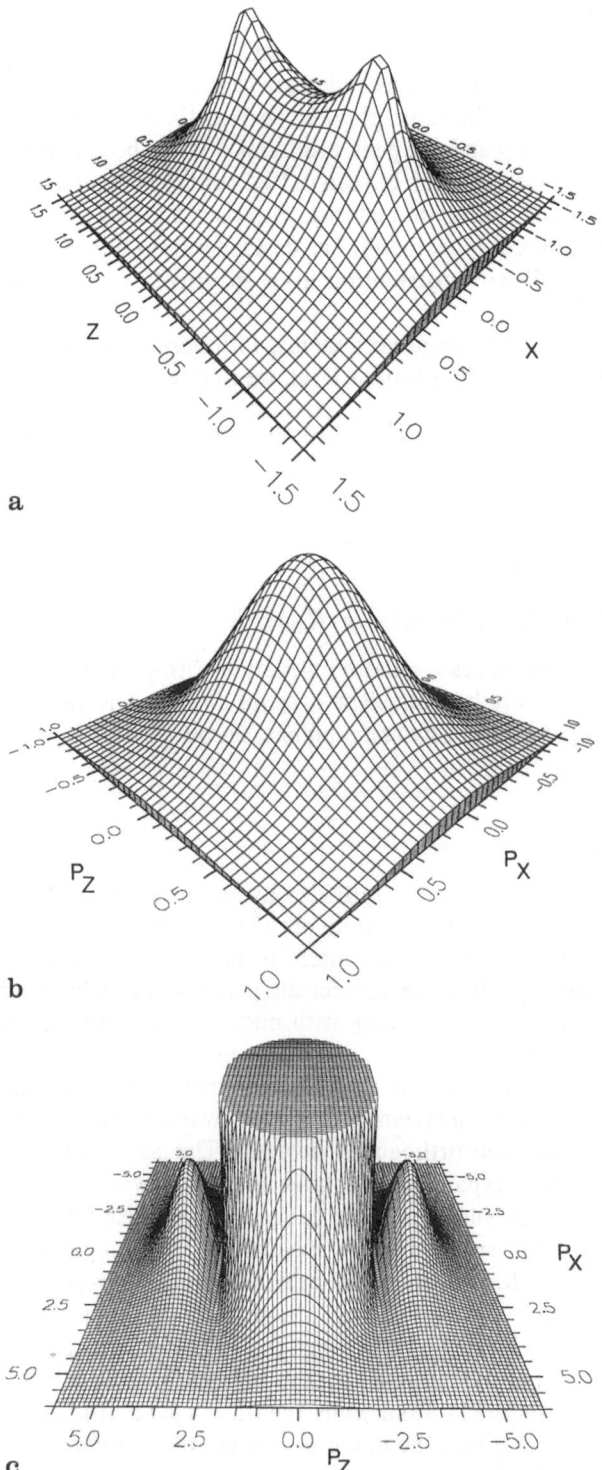

Note that $\rho(\boldsymbol{p}) = 0$ whenever $p_z R_{AB} = (2n + 1)\pi$, for integer values of n. Consequently, there is a series of *nodal planes* perpendicular to the p_z-axis. Figure 1c shows the momentum density in the same plane as in Fig. 1b, except that the view extends to larger values of p and the peak around $\boldsymbol{p} = 0$ has been omitted. The planes which correspond to $n = 0$ can be seen in Fig. 1c at $p_z = \pm \pi/R_{AB}$ (i.e. $p_z \approx \pm 2.2$ atomic units). These 'diffraction effects', with period $2\pi/R_{AB}$, contain the information about the molecular geometry. The cosine terms in Eqs. (14) and (15) change the momentum density distribution from spherical to ovaloid on bond formation.

The diffraction features for heteronuclear systems, with non-centrosymmetric MOs, are slightly different. At the simplest level, we may write

$$\psi(r) = c_A \phi_A(r - R_A) + c_B \phi_B(r - R_B) \tag{16}$$

so that

$$\begin{aligned}
\rho(\boldsymbol{p}) &= |\Psi(\boldsymbol{p})|^2 \\
&= c_A^2 |\Phi_A(\boldsymbol{p})|^2 + c_B^2 |\Phi_B(\boldsymbol{p})|^2 \\
&\quad + 2c_A c_B \mathrm{Re}\left[\Phi_A^*(\boldsymbol{p})\, \Phi_B(\boldsymbol{p}) \exp(-i\boldsymbol{p} \cdot \boldsymbol{R}_{AB})\right]. \tag{17}
\end{aligned}$$

As before, the exponential introduces oscillatory structure with period $2\pi/R_{AB}$. However, a key difference from homonuclear systems ($c_A = c_B$) is that the diffraction features no longer lead to any *nodal* planes, simply because $c_A \neq c_B$.

2.3 Bond Formation in BH$^+$

A further illustration of p-space concepts is provided by an investigation of the bond formation process in the $X^2\Sigma^+$ ground state of the BH$^+$ molecular ion (cf. [19–21]). It is, of course, important in this context to employ wavefunctions which are sufficiently flexible to allow for correct dissociation and which can describe properly the variation of the bonding with nuclear separation. In the particular case of BH$^+$ [19], we used a modern development of ab initio valence bond theory, known as the spin-coupled approach to molecular electronic structure [22], in which an N-electron system is described by a single configuration consisting of N distinct non-orthogonal orbitals. The so-called spin-coupled orbitals, $\phi_1 .. \phi_5$, were expanded in a large universal even-tempered basis set of Slater-type functions, such that each orbital could utilise functions stemming from either centre. The spin-coupled wavefunction was computed for numerous nuclear separations, R, with the bond oriented along the z-axis. We found, for all values of R, that ϕ_1 and ϕ_2 take the form of B(1s)-like 'core' orbitals. Orbitals ϕ_3 and ϕ_4 are based on B(2s) functions, but take on increasing $2p_z$ character for values of R approaching the equilibrium geometry, such that one orbital (ϕ_3) points towards the H atom while the other (ϕ_4) points away.

For very large nuclear separations, the mode of spin coupling is characteristic of the (correct) B$^+$ + H asymptote, in that the 'odd electron' occupies

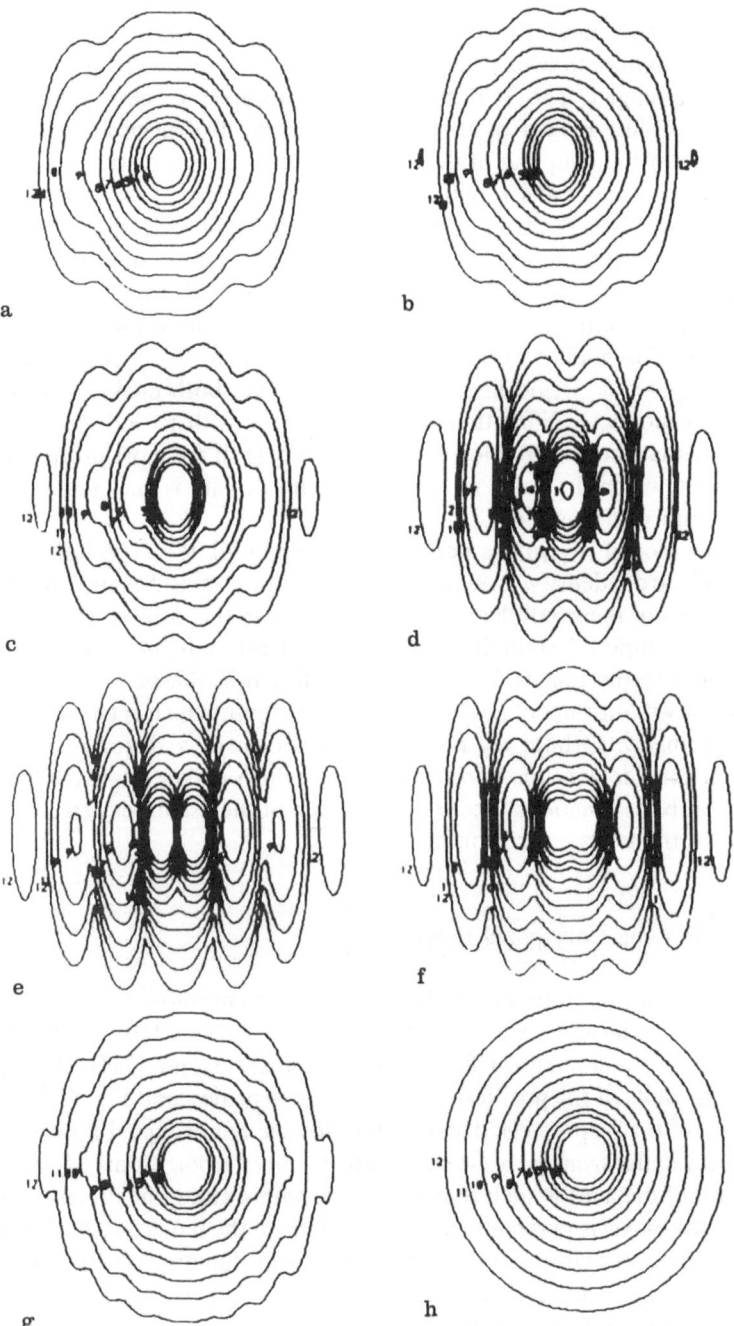

Fig. 2a–h. Contour plots of $|\phi_5(p)|^2$, taken from spin-coupled calculations on BH^+ ($X^2\Sigma^+$). The nuclear separations considered here are: **a** 2.3501 bohr; **b** 3.0 bohr; **c** 3.3 bohr; **d** 3.5 bohr; **e** 3.75 bohr; **f** 4.0 bohr; **g** 5.5 bohr; **h** 10.0 bohr. Each p-space electron density has cylindrical symmetry (about the p_z-axis)

orbital ϕ_5, which takes the form of an H(1s) function. At the equilibrium geometry, on the other hand, the dominant mode of spin coupling corresponds to a molecular situation, in which the electron spins associated with ϕ_3 and ϕ_4 are coupled to a singlet. The 'odd electron' now occupies orbital ϕ_4, which resembles a B(sp^x)-like hybrid pointing away from the hydrogen atom. This dramatic change in the coupling of the electron spins with decreasing R is far from gradual, and occurs most rapidly around 3–4 bohr.

The recoupling of the electron spins is accompanied by some distortion in orbital ϕ_5. These changes are particularly dramatic when viewed in p-space (see Fig. 2). As the two atoms approach one another, ϕ_5 starts to distort towards the boron cation. This results in some elongation of $|\phi_5(p)|^2$ along the p_z-axis, as shown in Fig. 2g. For smaller values of R, corresponding to the region of most rapid spin recoupling, ϕ_5 delocalises to some extent over both nuclear centres. This leads to pronounced diffraction effects in p-space with period $2\pi/R$ (see Fig. 2c–f). Note that these diffraction features (see Eq. (17)) are most prominent at values of p for which the value of the momentum density is very much less than its maximum value, as for H_2.

The contour plot of $|\phi_5(p)|^2$ is particularly striking for $R = 3.75$ bohr (see Fig. 2d). The electron density is elongated along the p_z-axis because of the angular properties of the small degree of p$_z$ character in this orbital. However, the plot is in fact composed of smaller ellipsoids which are elongated symmetrically *perpendicular* to the p_z-axis, because of the diffraction effects.

Figure 2a shows contours of $|\phi_5(p)|^2$ at $R \approx 2.35$ bohr, which is close to the equilibrium geometry. Orbital ϕ_5 now takes the form of a distorted H(1s) function which extends over the boron centre, but which makes relatively little use of the boron basis functions. Accordingly, much of the oscillatory structure disappears from the p-space electron density.

2.4 π-Orbitals in Long Conjugated Polyenes

Our final example in this section is the form of the momentum-space density in larger molecules, specifically large polyenes $H_2C{=}CH{-}(CH{=}CH)_m{-}CH{=}CH_2$ [23], which are also a convenient first model for polyacetylene $(CH)_x$. Figure 3 shows contour plots of the position-space electron densities, and Fig. 4 shows isometric views of the p-space electron densities for the ten occupied π-orbitals $(\pi_1 .. \pi_{10})$ of the regular *trans*-polyene containing twenty carbon atoms, $C_{20}H_{22}$. The orientation of the axes is such that the x-axis points along the direction of the chain, and the z-axis is perpendicular to the plane of the molecule, as shown in Fig. 3. These densities were calculated from SCF wavefunctions (3-21G basis set) generated using the GAMESS-UK program [15] (r(C–C) = 1.39 Å, r(C–H) = 1.09 Å, all bond angles 120°). The r-space electron densities, plotted in the plane $z = 1$ atomic unit, show the expected characteristic nodal structure. With decreasing binding energy, each successive π-orbital possesses one more nodal plane, such that orbital π_μ exhibits $\mu - 1$ nodes.

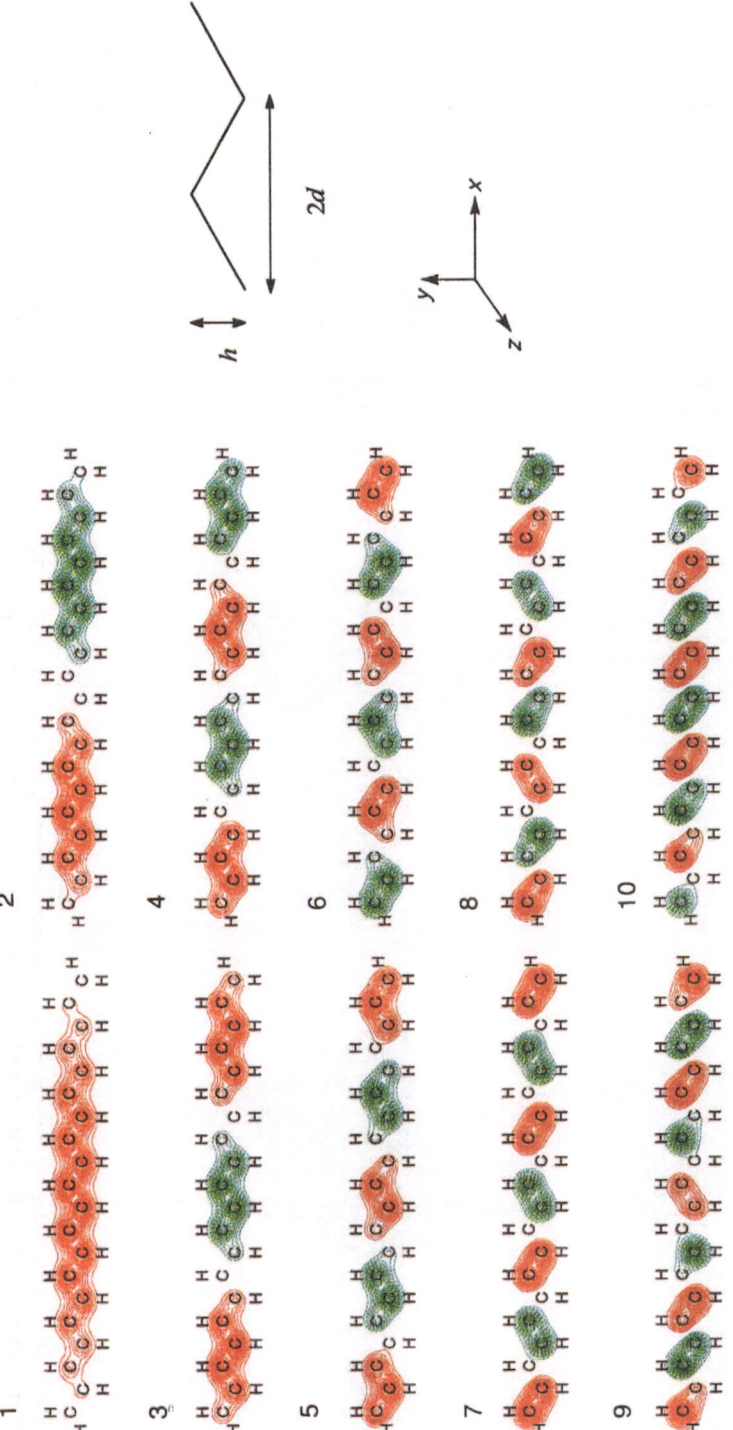

Fig. 3. Contour plots (in the plane $z = 1$ bohr) of the position-space electron density for the occupied π-orbitals for *trans*-$C_{20}H_{22}$. Red (*full*) and green (*broken*) contours denote regions in which the wavefunction has opposite phases. The labels mark the positions of the nuclei (projected onto the plane $z = 1$). Orbital π_1 has the highest binding energy and orbital π_{10} is the least strongly bound. Note also the orientation of the axes (the x-axis points along the chain and the z-axis perpendicular to the molecular plane), as well as the marked distances d and h to which reference is made in the text

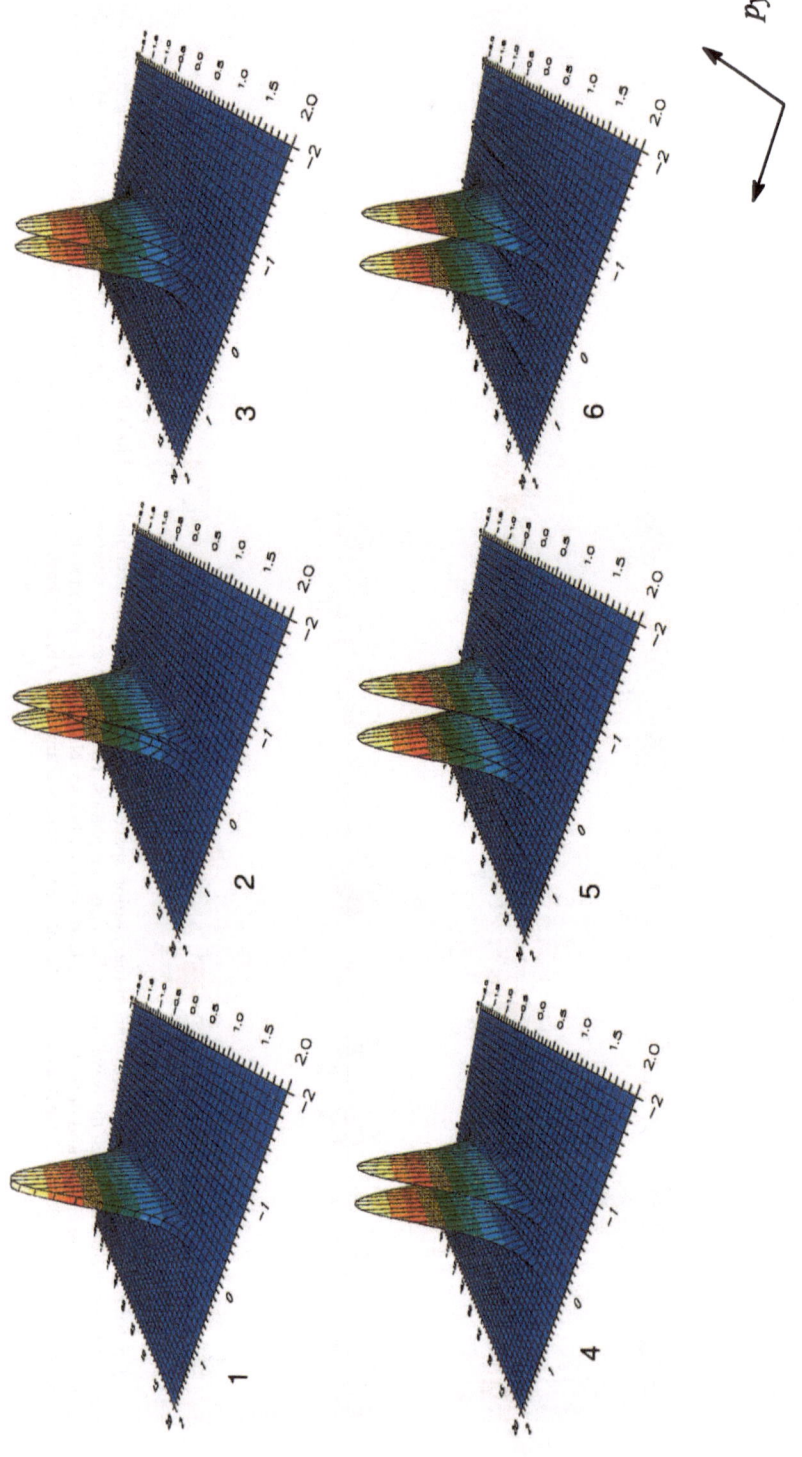

Fig. 4. p-space electron density in the plane $p_z = 1$ atomic unit for the orbitals shown in Fig. 3. The centre of each plot corresponds to the point $(0, 0, 1)$ atomic unit in momentum space. Each view extends to ± 2 atomic units in the p_x and p_y directions. The x, y and z labels refer to the same orientation of the axes as in Fig. 3

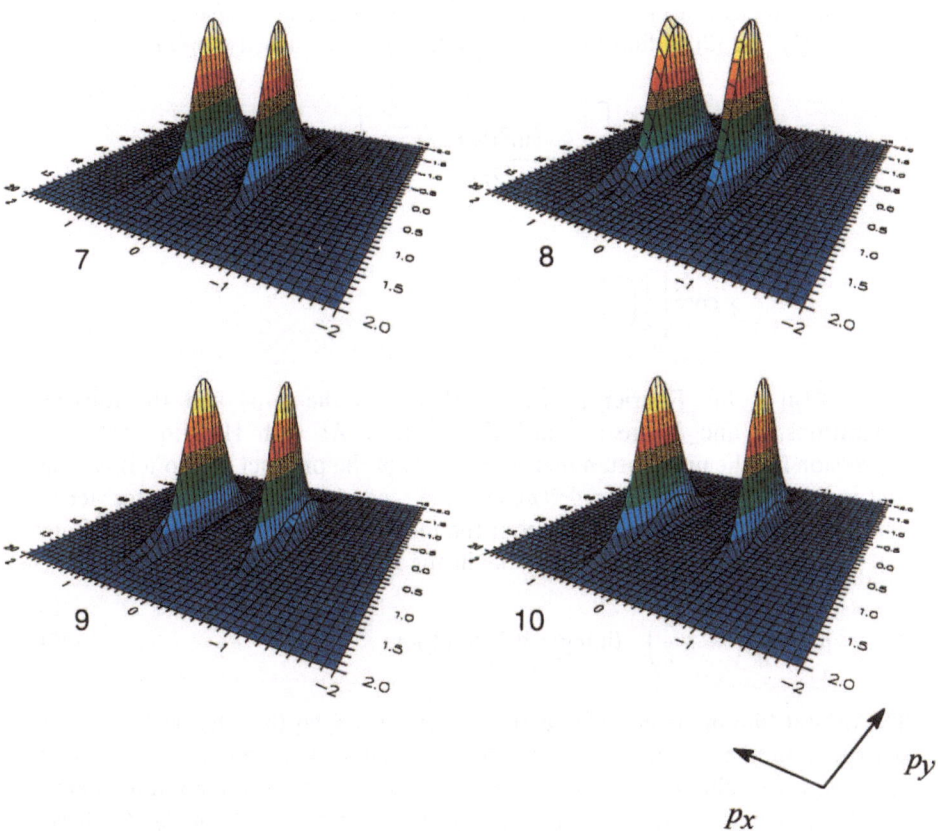

Fig. 4. Continued.

The p-space plots in Fig. 4 are very different in appearance from the r- and p-space plots for H_2 (see Fig. 1b, c). The momentum-space densities are plotted in the plane $p_z = 1$ atomic unit, with the *centre* of each view corresponding to $p_x = p_y = 0$. These plots show sharp wedges which, for each orbital, peak sharply at one particular value of $|p_x|$, i.e. at one value of the component of momentum *along* the chain direction. Only the π-orbital with the highest binding energy has non-zero momentum density at $p_x = 0$, and the wedges are furthest from the origin for the occupied π-orbital with the lowest binding energy (i.e. the highest occupied molecular orbital).

Simple Hückel theory can be used to rationalise these plots. In this approach, the MOs for a chain containing n carbon atoms are

$$\psi_k(r) = \frac{1}{\sqrt{n}} \sum_{j=1}^{n} \exp\left(\frac{2\pi ijk}{n}\right)\phi_j(r) \quad k = 0, 1, \ldots, n-1 \tag{18}$$

in which $\phi_j(r)$ is a normalised $2p_z$ atomic orbital on carbon atom number j. For simplicity, the overlap between nearest neighbours has been neglected. We can use Eqs. (1) and (5) to find the corresponding p-space density, $\rho_k(p)$:

$$\rho_k(p) = 4\Phi(p)\,\Phi^*(p)\left[\frac{\sin^2(\tfrac{1}{2}np_x d)}{n\sin^2\left(\dfrac{2\pi k}{n} - p_x d\right)}\right]$$

$$\times \cos^2\left[\frac{1}{2}\left(\frac{2\pi k}{n} - p_x d - p_y h\right)\right] \tag{19}$$

where $\Phi(p)$ is the Fourier transform of *any* of the $\phi_j(r)$ and the nuclear separations, d and h, are as marked in Fig. 3. As with H_2 (Eq. (13)). the expression for the momentum density consists of the product of two terms. The first is an 'atomic term', $\Phi(p)\Phi^*(p)$, and the second a 'diffraction factor' term, which contains the information about the molecular geometry. It follows from Eq. (19) that there will be sharp peaks in the momentum density whenever

$$p_x = \frac{2\pi}{d}\left(m + \frac{k}{n}\right) \quad \text{(integer values of } m\text{)}. \tag{20}$$

The orbital binding energy decreases as k increases, so that the peaks for less strongly bound electrons (larger k) occur at successively larger values of p_x. For any non-zero value of p_z, only the π-orbital with the largest binding energy ($k = 0$) has a maximum at $p_x = p_y = 0$. This is all shown clearly in Fig. 4, where, for each orbital, we can see the peaks corresponding to $m = 0$. The highest occupied molecular orbital (HOMO, $k = 9$), for example, has a sharp peak, as expected, at $p_x = 0.9\pi/d \approx 0.7$ atomic units. Further peaks for $m \neq 0$ will occur at larger values of p_x. From Eq. (20), the repeat distance is $2\pi/d$. These peaks will be much smaller, because the atomic term falls off so rapidly with p (in the present case as p^{12}).

With increasing n, the peaks in the momentum density become more marked. This is as expected for delocalised bonding, since r and p are conjugate variables. In the limit $n \to \infty$, $\rho_k(p)$ is *only* non-zero when Eq. (20) is satisfied. The diffraction term in Eq. (20) restricts the range of allowed momenta in the MOs much more than in those of smaller molecules, such as H_2.

The momentum-space representation also proves particularly convenient for comparisons of the electron distributions of systems with different nuclear frameworks. 'Difference density plots' in r-space are complicated by the different sets of nuclear positions. Such complications are *absent* in p-space and, in the case of polyenes [23], for example, momentum-space concepts have proved useful for examining the effects of bond alternation on the electron density – an important characteristic of such systems and of doped polyacetylene.

3 Molecular Similarity and Dissimilarity Indices: Definitions

We start by introducing the quantity central to our approach, namely the generalised overlap $I_{AB}(n)$, defined according to

$$I_{AB}(n) = \int p^n \rho_A(p) \rho_B(p) \, dp \qquad (21)$$

in which $\rho_A(p)$ and $\rho_B(p)$ are the momentum-space electron densities of molecules A and B. These momentum densities can be *total* electron densities, *total valence* electron densities, or the densities associated with one or more individual *orbitals*. It is possible to compute $I_{AB}(n)$ for particular orbitals of two different molecules (e.g. their respective HOMOs or localised orbitals describing particular bonds), or the densities associated with a particular atom or group of atoms common to the two molecules. A further option is to consider two orbitals of the *same* molecule, such as the HOMO and the lowest unoccupied molecular orbital (LUMO), for example. The purpose of introducing the term p^n into Eq. (21) is to weight preferentially different regions of p-space (cf. the moments of momentum defined in Eq. (7)). Typical values of n are again $-1, 0, 1$, and 2. For example, a factor of p^{-1} in the integrand places particular emphasis on the slowest moving electrons; this corresponds in r-space to highlighting the long-range slowly-varying outer valence electron density.

Values of $I_{AB}(n)$ depend on the relative orientation of the two molecules. However, $\rho_A(p)$ and $\rho_B(p)$ are independent of the choice of origin in r-space, and so $I_{AB}(n)$ is completely independent of the separation between the two systems in position space. Of course, the relationship between r-space and p-space is such that it would be possible to transform Eq. (21) into an equivalent position-space integral. In practice, however, it is *much* easier to work directly in p-space with Eq. (21).

$I_{AB}(n)$ is currently computed by direct numerical evaluation, typically using standard library routines. The most appropriate method of integration varies from case to case. Three such schemes are currently in use – all of them yield exactly the same results, but the computational effort can vary significantly, depending on the characteristics of the integrand. The first of these methods uses an adaptive strategy with repeated subdivision of a cube into smaller three-dimensional boxes. In each subregion, the integrand is estimated using a seventh degree rule and the 'error' is estimated by comparison with a fifth degree rule. Hyper-rectangular subregions associated with large errors are subdivided further. The second of our schemes is much the same as the one just described, except that the range of integration is a sphere rather than a cube, and the various subregions are portions of spherical shells rather than of boxes. We have found that this approach tends to be particularly efficient when dealing with total densities for molecules or for parts of molecules. The third of our algorithms integrates over a cube using the method of Sag and Szekeres [24], which exploits various properties of the shifted p-point trapezoidal rule. This scheme tends to be much more efficient than the other two when dealing with limited summations over individual orbitals. For $n = -1$, it is usually sufficient to integrate out to $p_{max} = 5$ atomic units.

The integrals $I_{AB}(n)$ can take any non-negative value, and so it is natural to attempt some form of normalisation. The first strategy we used, following work by Carbó [8], was to divide $I_{AB}(n)$ by the geometric mean of the 'self' terms:

$$R_{AB}(n) = \frac{100 \times I_{AB}(n)}{\sqrt{I_{AA}(n) \times I_{BB}(n)}}. \tag{22}$$

It follows that the similarity indices $R_{AB}(n)$ must lie in the range 0–100%, with higher values implying greater similarity. An obvious alternative to scaling $I_{AB}(n)$ by the geometric mean of $I_{AA}(n)$ and $I_{BB}(n)$ is to use instead the arithmetic mean:

$$S_{AB}(n) = \frac{100 \times I_{AB}(n)}{\frac{1}{2}(I_{AA}(n) + I_{BB}(n))}. \tag{23}$$

The similarity indices $S_{AB}(n)$ are also in the range 0–100%. Higher values imply greater similarity. An expression of this type was first discussed in detail by Hodgkin and Richards [10]. If $\rho_A(p) = m \times \rho_B(p)$, $R_{AB}(n)$ is invariant to the choice of (non-zero) m, whereas $S_{AB}(n)$ is not. In this sense, values of $S_{AB}(n)$ are less dominated by the *shape* of the p-space electron densities than are the values of $R_{AB}(n)$. This tends to make the Hodgkin-like definition more appropriate than the Carbó-like definition for some p-space applications.

A third definition, with a number of practical advantages, was suggested by a quantity utilised in statistical analysis. Our Tanimoto-like similarity index takes the form:

$$T_{AB}(n) = \frac{100 \times I_{AB}(n)}{I_{AA}(n) + I_{BB}(n) - I_{AB}(n)}. \tag{24}$$

Like the other indices, $T_{AB}(n)$ can adopt values in the range 0–100%, with higher values implying greater similarity. Of course, it is straightforward to show that the indices $S_{AB}(n)$ and $T_{AB}(n)$ are not independent:

$$\frac{100}{T_{AB}(n)} = \frac{200}{S_{AB}(n)} - 1. \tag{25}$$

The variation of $T_{AB}(n)$ with $S_{AB}(n)$ is illustrated in Fig. 5, from which it is evident that the Tanimoto-like index is somewhat more discriminating than the Hodgkin-like index when dealing with high values (i.e. with very similar systems). Of course, this conclusion is *not* specific to p-space indices.

When dealing with quantities which show very high similarity, such as the p-space electron densities for analogous bonds in closely related molecules, we have found it more appropriate to dispense with the normalisation of $I_{AB}(n)$. A much more discriminating definition for such cases is the family of p-space

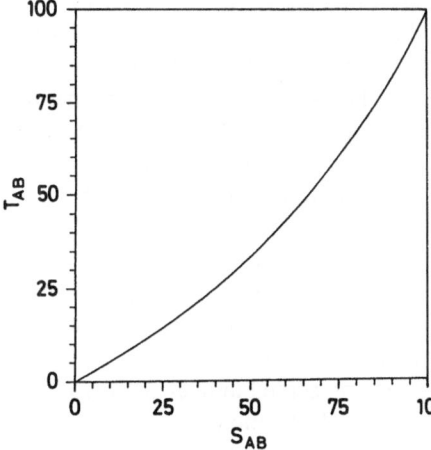

Fig. 5. Relationship between values of T_{AB} and S_{AB} (Eq. (25))

dissimilarity indices:

$$D_{AB}(n) = 100 \times (I_{AA}(n) + I_{BB}(n) - 2I_{AB}(n)) \tag{26}$$

in which the ratio in $S_{AB}(n)$ has been replaced by a difference. This distance-like index cannot be negative and larger values of $D_{AB}(n)$ imply greater dissimilarity, without an upper bound. The magnitude of $D_{AB}(n)$ scales to a larger extent with $I_{AB}(n)$, so that it should normally only be employed when $\rho_A(\boldsymbol{p})$ and $\rho_B(\boldsymbol{p})$ are very similar indeed, i.e. when they correspond to systems that are extremely closely related. This last observation is self-evident, given the relationships linking $D_{AB}(n)$ with the normalised quantities $S_{AB}(n)$ and $T_{AB}(n)$:

$$\frac{D_{AB}(n)}{100} = I_{AB} \times \left(\frac{100}{T_{AB}(n)} - 1 \right)$$

$$= 2 \times I_{AB}(n) \times \left(\frac{100}{S_{AB}(n)} - 1 \right). \tag{27}$$

4 Model Systems

4.1 CH_3OCH_3, CH_3SCH_3 and $CH_3CH_2CH_3$

We begin by comparing the compounds CH_3OCH_3, CH_3SCH_3 and $CH_3CH_2CH_3$, which have been examined in numerous similarity studies [10, 12–14]. Comparable biological activity is often obtained when a $-CH_2-$ group in an aliphatic chain is replaced by $-S-$. The same is not usually the case when $-O-$ is substituted for $-CH_2-$.

Table 2 lists values of $R_{AB}(-1)$, $S_{AB}(-1)$ and $T_{AB}(-1)$ calculated using total p-space densities computed from ab initio SCF wavefunctions (4-31G basis). It is straightforward to choose an appropriate relative orientation for this series of molecules: the z-direction is the C_2-axis and the y-axis lies in the σ'_v plane. R_{AB} and S_{AB} are tabulated in [12] for the additional cases $n = 0, 1$ and 2. The same qualitative variation is observed from one molecule to another, but the range is smaller. For each value of n, the largest similarity indices are between $CH_3CH_2CH_3$ and CH_3SCH_3, and the smallest between CH_3SCH_3 and CH_3OCH_3, consistent with the relative biological activity generally observed. The difference between the largest and the smallest values of $R_{AB}(-1)$, which depends particularly on the *shapes* of the two electron densities, is only 0.1. S_{AB} is less dominated by considerations of shape, but the range of values of $S_{AB}(-1)$ is still small (2.4). As expected from the argument presented in Sect. 3, the range of $T_{AB}(-1)$ is larger (4.6). For these systems, where all the molecules have much in common, none of the p-space similarity indices is particularly discriminating.

Values of $D_{AB}(-1)$ for this model series, computed using the same ab initio SCF wavefunctions, are also listed in Table 2. The same conclusions are obtained from this distance-like dissimilarity index as with the three similarity indices, but there is a very much larger variation (45.4). We have found that D_{AB} is most useful in situations in which all the molecules are very closely related. There is no upper limit to $D_{AB}(n)$.

The calculation of ab initio wavefunctions for the larger molecules associated with applications such as those discussed later in this paper is computationally demanding. Accordingly, we have also examined these model systems using semi-empirical wavefunctions taken from the well-established MOPAC program [16]: such wavefunctions can be obtained very cheaply. Table 3 collects together values of R_{AB}, S_{AB}, T_{AB} and D_{AB} for the model series, evaluated using

Table 2. Values of the similarity indices $R_{AB}(-1)$, $S_{AB}(-1)$, and $T_{AB}(-1)$ and the dissimilarity index $D_{AB}(-1)$ for the model series, $(CH_3)_2X$ ($X = CH_2$, S or O), calculated using ab initio wavefunctions

	$R_{AB}(-1)$	$S_{AB}(-1)$	$T_{AB}(-1)$	$D_{AB}(-1)$
$CH_3CH_2CH_3/CH_3OCH_3$	99.8	98.4	96.8	29.4
$CH_3CH_2CH_3/CH_3SCH_3$	99.9	99.8	99.6	4.0
CH_3OCH_3/CH_3SCH_3	99.7	97.4	95.0	49.4

Table 3. Values of the similarity indices $R_{AB}(-1)$, $S_{AB}(-1)$, and $T_{AB}(-1)$ and the dissimilarity index $D_{AB}(-1)$ for the model series, calculated using MNDO wavefunctions

	$R_{AB}(-1)$	$S_{AB}(-1)$	$T_{AB}(-1)$	$D_{AB}(-1)$
$CH_3CH_2CH_3/CH_3OCH_3$	99.8	97.6	95.4	137
$CH_3CH_2CH_3/CH_3SCH_3$	99.9	99.9	99.9	5
CH_3OCH_3/CH_3SCH_3	99.6	97.3	94.7	161

Table 4. Values of the similarity indices $R_{AB}(-1)$, $S_{AB}(-1)$, and $T_{AB}(-1)$ and the dissimilarity index $D_{AB}(-1)$, once more for the model series (using MNDO wavefunctions), but ignoring the contribution to the electron density from the methyl groups

	$R_{AB}(-1)$	$S_{AB}(-1)$	$T_{AB}(-1)$	$D_{AB}(-1)$
$CH_3CH_2CH_3/CH_3OCH_3$	88.2	56.2	39.0	170
$CH_3CH_2CH_3/CH_3SCH_3$	98.9	98.7	97.5	8
CH_3OCH_3/CH_3SCH_3	90.9	60.8	43.7	136

MNDO wavefunctions. Again, the conclusions regarding the relative similarities are unchanged, with CH_3SCH_3 and $CH_3CH_2CH_3$ the most similar pair. It is particularly useful that the same trends are evident in the indices whether the electron densities are calculated from ab initio or semi-empirical techniques. This observation offers considerable justification for the use of the cheap, semi-empirical methods when working with series of large molecules. Of course, the ab initio and semi-empirical numbers are not the same, and would not be expected to be so. This simply reflects the different qualities of the two types of wavefunction. In addition, the MOPAC calculations generate 'valence-electron' densities whereas the ab initio densities are 'all-electron'. This last effect should be small, because I_{AB} emphasises the valence region of the electron density, for the reasons discussed in Sect. 3.

It is easy to envisage situations in which it would be more appropriate to compare only certain parts of a series of molecules, such as fragments containing the active centres. One way of proceeding is simply to compare the electron densities associated with the atoms of interest. This requires the use of some 'atoms-in-molecules' method, by which the r-space electron density is assigned to individual atoms. For example, Cioslowski and Nanayakkara [25], working with similarity measures based on r-space electron densities, have chosen the zero-flux partitioning scheme of Bader [26]. We have implemented a much more primitive, and considerably cheaper, method: we consider only the basis functions centred on the atom(s) of interest. Returning to the model series, it is simple to compare solely those parts of the molecules that are not in common, i.e. the $-O-$, $-S-$ and $-CH_2-$, linkages. Table 4 lists the values of $R_{AB}(-1)$, $S_{AB}(-1)$, $D_{AB}(-1)$ and $T_{AB}(-1)$ obtained in this way, using MNDO wavefunctions. The same pair of molecules, i.e. CH_3SCH_3 and $CH_3CH_2CH_3$, is once again the most similar, although the relative order of the other two pairs is now reversed. As might have been anticipated, the range of values of each index is somewhat larger than in Table 3.

4.2 Hydrofluoromethanes $CH_{4-x}F_x$

An attractive alternative to the strategy described at the end of Sect. 4.1 is the use of a reliable and computationally cheap scheme to localise the occupied molecular orbitals. This has enabled us to compare the C–H and C–F bonds in the hydrofluoromethane molecules, CH_xF_{4-x} $(0 \leq x \leq 4)$ (also considered by Cioslowski and Nanayakkara [25]). The AM1 semi-empirical method [16] was used here instead of MNDO, because the latter tends to produce somewhat larger errors for highly fluorinated compounds [27]. AM1 wavefunctions were generated with the z-axis lying along a C–F bond and localised molecular orbitals were then produced by the Mulliken population procedure introduced by Pipek and Mezey [28]. In this approach, the quantity

$$Z = \sum_A \sum_\mu |P_A(\mu)|^2 \tag{28}$$

is maximised, in which $P_A(\mu)$ is the contribution from each electron in MO ψ_μ to the Mulliken population on centre A.

In a different context, we have considered how primary C–F bonds in fluoroalkanes differ very little from molecule to molecule [29]. The C–F bond lengths and the Mulliken charges on F, calculated by the AM1 method, vary very little from system to system. The charge distribution in CF_4 is the only slight exception, in that the Mulliken charge on F is a little less negative than in the other molecules. Table 5 lists values of $D_{AB}(-1)$ for the localised orbitals that describe the C–F bonds. All these bonds are very similar. Values of $D_{AB}(-1)$ for individual orbitals are always smaller than those for the corresponding total densities, simply because this index involves a difference and not a ratio. Nevertheless, it is clear that the C–F bonds in 'consecutive' molecules in the series CH_3F, CH_2F_2, CHF_3 and CF_4 are the most similar. The values of $D_{AB}(-1)$ are slightly larger when CF_4 is involved in the comparison. It is evident that $D_{AB}(-1)$ can differentiate between the C–F bonds in this series.

In order to compare the C–H bonds, AM1 wavefunctions were calculated for methane and for the hydrofluoromethanes, orientating each molecule such that a C–H bond lay along the z-axis. Localised molecular orbitals were produced using the population localisation method and values of $D_{AB}(-1)$ calculated (see Table 6). The AM1 calculations reproduce the experimental observation that the C–H bond length is insensitive to increasing fluorination. Nevertheless, the C–H bonds in these molecules do show a marked variation in chemical reactivity. This is evident from the rate constants and from the activation energies for H-atom abstraction by free radicals. The reaction with OH is of particular

Table 5. Values of the dissimilarity index $D_{AB}(-1)$ for the C–F bonds in the series $CH_{4-x}F_x$, $(x = 1, 2, 3, 4)$

	CH_2F_2	CHF_3	CF_4
CH_3F	0.2	0.4	1.5
CH_2F_2		0.2	1.2
CHF_3			0.8

Table 6. Values of the dissimilarity index $D_{AB}(-1)$ for the C–H bonds in the series $CH_{4-x}F_x$, $(x = 0, 1, 2, 3)$

	CH_3F	CH_2F_2	CHF_3
CH_4	1.1	3.9	13.1
CH_3F		1.9	9.2
CH_2F_2			4.5

significance, because such processes constitute the dominant loss process for hydrofluorocarbons in the troposphere. Several models [30, 31] predict that the tropospheric lifetime of CHF_3 is much longer than that of CH_2F_2 or of CH_3F. If we assume a correlation between the properties of the molecules themselves and those of the corresponding transition states, then we would anticipate that the $D_{AB}(-1)$ values for the C–H bonds should be larger than those for the C–F bonds. This is evident from a comparison of Tables 5 and 6. The largest value of $D_{AB}(-1)$ is that for CHF_3. The rate constant of the reaction with the OH radical at 298 K is much smaller, and the activation energy much greater, for CHF_3 than for any of the other hydrofluoromethanes [32]. This much lower reactivity of CHF_3 is highlighted by the p-space dissimilarities calculated using localised molecular orbitals.

5 Anti-HIV1 Phospholipids

Our next example [33] relates to a series of larger molecules, namely the virology data for a particular group of phospholipids [34, 35]. Those considered here have the general structural formula

in which R^2 represents the lipid tail and R^1 is a relatively small alkyl substituent on the amine group. Molecules with different choices of R^2 and R^1 give 50% inhibition of HIV1 in C8166 T-lymphoblastoid cells with the ED_{50} (μM) values listed in Table 7. There appears to be no *obvious* correlation between HIV1 inhibition and the identity of the lipid tail or, indeed, of other alkyl substituents on the amine group. We have been unable to make much sense of the various trends by consideration of chemical structure alone. For example, the replacement by t-butyl of the methyl group at R^1 results in greatly reduced ED_{50} values

Table 7. ED_{50} values for the inhibition of HIV1 in C8166 T-lymphoblastoid cells and two calculated similarity indices: $T_{AB}^{(HX1)}$ is the similarity of the given molecule to the *inactive* compound HX1 and $T_{AB}^{(OL2)}$ is the similarity to the *active* compound OL2

Mnemonic	ED_{50} (μM)	R^1	R^2	$T_{AB}^{(HX1)}$	$T_{AB}^{(OL2)}$
HX1	> 200	methyl	n-hexyl	100	61.0
DD1	> 200	methyl	n-dodecyl	86.5	87.0
OD1	25	methyl	n-octadecyl	67.6	98.9
EG1	110	methyl	ethyl glycolate	97.4	51.4
OL1	10	methyl	oleyl	68.8	98.4
HX2	40	t-butyl	n-hexyl	95.4	75.5
DD2	10	t-butyl	n-dodecyl	76.6	94.8
OD2	3	t-butyl	n-octadecyl	59.9	99.9
EG2	200	t-butyl	ethyl glycolate	99.0	67.0
OL2	3	t-butyl	oleyl	61.0	100
HX3	> 200	hydrogen	n-hexyl	99.2	55.3
DD3	4	hydrogen	n-dodecyl	89.6	83.2
OD3	3.5	hydrogen	n-octadecyl	70.6	97.6
OL3	0.5	hydrogen	oleyl	71.7	97.4

in the most inactive compounds (HX1 and DD1), and in some enhancement in the activity of two others (OD1 and OL1), whereas it is clear from the data in Table 7 that EG 2 is *less* active than EG1. The series of compounds in which the amine group is free (R^1 = H) tends to have very low ED_{50} values, except that HX3 is *inactive*, unlike HX2.

The mechanism by which these phospholipids inhibit the virus is not completely understood, although it is thought that they first insert into the membranes of the virus. The concept of molecular similarity is particularly appropriate here, because the molecular processes involved are so complex. Bearing in mind that it is likely that the phospholipids must be able to mimic the membrane lipids, we chose to compare total densities for the complete molecules, rather than the densities for individual orbitals or for molecular fragments.

In view of the size of the phospholipids, computationally inexpensive r-space wavefunctions were generated from semi-empirical (MNDO) geometry optimisations. $T_{AB}(-1)$ was calculated for each pair of molecules, with the phospholipids aligned so as to match as closely as possible the positions of those atoms common to all the systems. The values are listed in Table 7. Results are presented here only for the zwitterions, although it should be stressed that all the key features were also exhibited when the neutral molecules were compared.

We display in Fig. 6 a representative subset of our results – comparisons relative to the *inactive* compound HX1 and comparisons relative to the highly *active* compound OL2. The same trends were apparent for all pairs of molecules. The corresponding ED_{50} values from Table 7 are also shown, with an arbitrary value of 200 μM assigned to the most inactive compounds (HX1, DD1 and HX3). It is clear from the figure that active compounds have a high similarity to OL2 and a low similarity to HX1, and vice versa. The way in which the various maxima and minima coincide is particularly striking in both plots.

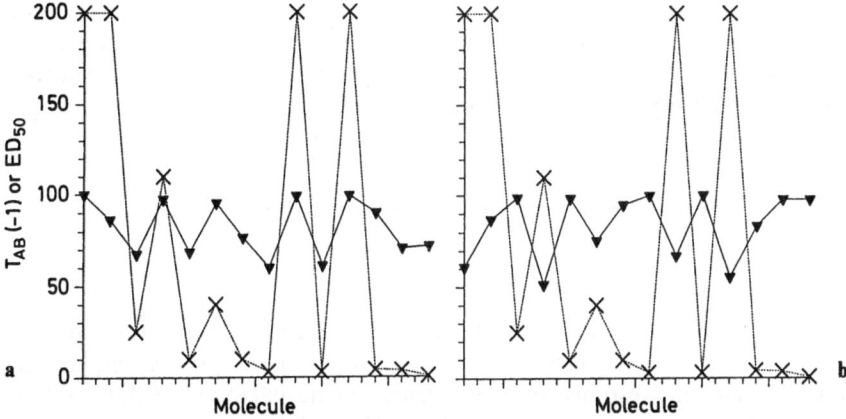

Fig. 6. Values of the molecular similarity index $T_{AB}(-1)$ ($\blacktriangledown$) for various phospholipids relative to: **a** the *inactive* compound HX1; **b** the highly *active* compound OL2. Also shown in each frame are the corresponding ED_{50} (μM) values ($\times$), with an arbitrary value of 200 μM assigned to the most inactive compounds (HX1, DD1 and HX3). All the values are taken from Table 7

We were initially informed that OD2 was insoluble/inactive, whereas we expected from the various $T_{AB}(-1)$ values that this phospholipid should have a very low ED_{50} value. The most recent data (Table 7) suggest $ED_{50} = 3$ μM. It is also worth pointing out that the T_{AB} values for the series with a free amine group were computed *before* the virology data became available to us. In this way we were able to *predict* that OD3 and OL3 would both have low ED_{50} values, but that HX3 would be inactive. For DD3 we observed moderately high T_{AB} values relative to *both* HX1 and OL2, and so it was impossible to make a definitive prediction in this case.

It is clear that the particular approach to molecular similarity employed here can be used to rationalise HIV1 virology data for the families of phospholipids considered and even to make some successful predictions of active compounds. Preliminary results for various AZT derivatives (reverse transcriptase inhibitors) are also encouraging. The active compounds all feature an $-N_3$ group. Nevertheless, comparisons of the HOMOs, which are localised in the thymine rings, can be used to distinguish easily between active and inactive compounds. Similarly, it has proved possible to identify features that are common to various, structurally-dissimilar, non-nucleoside reverse transcriptase inhibitors.

6 Hyperpolarisabilities of 1,4-Substituted Benzene Derivatives

We conclude with a non-biological example: materials with non-linear optical properties. We concentrate on structure-activity relations involving the hyperpolarisabilities, β, of conjugated molecules. The *direct* calculation of such hyperpolarisabilities is not only very difficult, but also extremely sensitive to the quality of the wavefunction. An alternative approach is *not* to attempt to calculate β a priori, but instead to look for empirical correlations involving β and some quantities which can be evaluated much more easily. Of course, it is important that such quantities should be relatively insensitive to the quality of the wavefunction.

The molecules of interest here can be represented as

D–(π)–**A**

in which –(π)– denotes a conjugated π-electron system (cf. the polyenes in Sect. 2.4), and **D** and **A** are electron-donor and electron-acceptor groups, respectively. Conventionally, the response of such systems to external electric fields is rationalised by invoking a two-state model [36, 37] involving

D$^+$–(π)–**A**$^-$.

At the simplest level, this second state can be envisaged as arising via excitation of an electron from the HOMO to the LUMO. Hence, it is natural to compare the HOMOs and LUMOs of such systems in *p*-space.

Results for one family of molecules are presented here. These are all 1,4-substituted benzene derivatives of the type

D–◯–A

All the required wavefunctions were generated using MNDO geometry optimisations. However, the calculated MNDO geometries for a few molecules exhibited some serious errors. For example, MNDO predicts, incorrectly, that the coordination at the nitrogen atom in Ph–NMe$_2$ is pyramidal, rather than planar. In such cases, appropriate experimental values were substituted for the relevant parameters.

For *each* molecule we have calculated the similarity index $R_{AB}(-1)$ between its HOMO and its LUMO, considering only contributions from basis functions centred on atoms in the benzene ring (the structural unit common to all the molecules). Accordingly, the problem of relative orientation does not arise. Figure 7 shows a plot of the experimental values of β vs $R_{AB}(-1)$. The experimental values were taken from [36] – all the values are listed in Table 8. It is clear that there is a reasonable correlation which could now be used predictively.

Further series of non-centrosymmetric conjugated molecules are currently under investigation [38]. In keeping with the known failure [36] of the two-state model for the family of stilbenes, Ar^1–CH=CH–Ar^2, our results indicate that such systems do not follow the same correlation observed for the benzene derivatives.

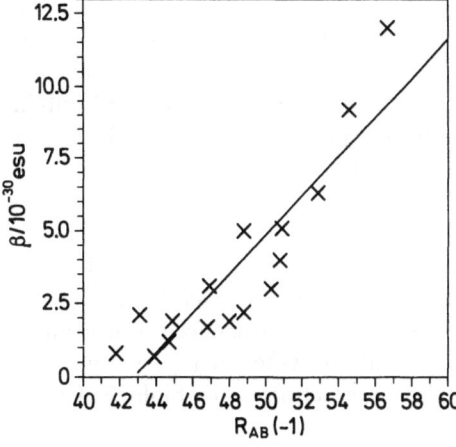

Fig. 7. Experimental values of the hyperpolarisability β (10^{-30} esu) vs the calculated orbital similarity $R_{AB}(-1)$ for a series of 1,4-substituted benzene derivatives. All the data are listed in Table 8

Table 8. Hyperpolarisabilities β (10^{-30} esu), taken from experiment [36], and values of the calculated orbital similarity index $R_{AB}(-1)$ for a series of non-centrosymmetric 1,4-substituted benzene derivatives

Acceptor	Donor	$R_{AB}(-1)$	β (10^{-30} esu)
CN	Cl	41.8	0.8
CN	Me	43.9	0.7
CN	OPh	44.7	1.2
CN	OMe	44.9	1.9
CN	NH_2	46.9	3.1
CN	NMe_2	48.8	5.0
COH	Me	46.8	1.7
COH	OPh	48.0	1.9
COH	OMe	48.4	2.2
COH	NMe_2	52.9	6.3
NO_2	Me	43.1	2.1
NO_2	OH	50.3	3.0
NO_2	OPh	50.8	4.0
NO_2	OMe	50.9	5.1
NO_2	NH_2	54.6	9.2
NO_2	NMe_2	56.7	12

7 Discussion

Momentum-space similarity and dissimilarity indices tend to be especially useful in situations for which the bonding topology is less important than the long-range valence electron density. In addition to model systems, the approach has now been applied successfully to a number of 'real' problems, such as anti-HIV1 phospholipids, reverse transcriptase inhibitors, and conjugated systems with high hyperpolarisabilities. In general, the method can be particularly effective for situations in which conventional chemical intuition is insufficient.

In addition to the choice of relative orientation, conformational flexibility may prove to be more of a problem in future applications than for the systems described here. This is particularly likely when the molecules are *very* similar. Of course, it is not hard to anticipate situations for which a *p*-space approach is unlikely to be useful. Obvious examples are those in which the observed activity is largely dependent on the values of bulk physical properties or on well-defined steric effects. We have stressed previously [12] that no *single* approach to molecular similarity can be expected to be a universal panacea. For example, any comparative methodology, such as that presented here, can only very indirectly provide information concerning interactions with biological receptors.

In summary, our unconventional approach to molecular similarity, based on indices derived from *momentum-space* electron densities, appears to show considerable promise for a wide range of applications. Work on a wide range of applications, both biological and non-biological, is currently in progress and the results will be reported in due course.

8 References

1. Williams BG (ed) (1977) Compton scattering. McGraw-Hill, New York
2. See for example: Brion CE (1986) Int J Quantum Chem 29: 1397
3. Coulson CA, Duncanson WE (1941) Proc Camb Philos Soc 37: 55, 67, 74, 397, 406
4. See for example: Epstein IR, Tanner AC (1977) In: Williams, New York; Rawlings DC, Davidson ER (1985) J Phys Chem 89: 969 and references therein
5. Johnson MA, Maggiora GM (eds) (1990) Concepts and applications of molecular similarity. Wiley, New York
6. Proceedings from the 1992 Beilstein symposium on similarity in organic chemistry (1992) J Chem Inf Comput Sci 32, no 6
7. Carbó R, Mezey PG (eds) Molecular similarity and reactivity: From quantum chemical to phenomenological approaches. Kluwer (in press)
8. See for example: Carbó R, Leyda L and Arnau M (1980) Int J Quant Chem 17: 1185; Carbó R, Domingo Ll (1987) Int J Quant Chem 32: 517; Carbó R, Calabuig B (1992) Int J Quant Chem 42: 1681, 1695
9. Ponec R, Strnad M (1991) J Phys Org Chem 4: 701; (1992) Int J Quant Chem 42: 501
10. Hodgkin EE, Richards WG (1987) Int J Quantum Chem Quantum Biology Symp 14: 105; Richards WG, Hodgkin EE (1988) Chem Br 24: 1141; see also Burt C, Richards WG (1990) J Comput-Aided Mol Design 4: 23
11. Duane-Walker P, Artera GA, Mezey PG (1991) J Comput Chem 12: 220
12. Cooper DL, Allan NL (1989) J Comput-Aided Mol Des 3: 253
13. Cooper DL, Allan NL (1992) J Am Chem Soc 114: 4774
14. Allan NL, Cooper DL (1992) in Ref 6, p 587
15. Guest MF, Sherwood P (1992) GAMESS-UK User's guide and reference manual, revision B.0; SERC Daresbury Laboratory, UK
16. Stewart JJP (1990) J Comput-Aided Mol Des 4: 1
17. Kaijser P, Smith VH Jr (1977) Adv Quant Chem 10: 37

18. See for example: Allan NL, Cooper DL (1986) J Chem Phys 84: 5594
19. Cooper DL, Allan NL (1987) J Chem Soc, Faraday Trans 2, 83: 449
20. Allan NL, Cooper DL (1987) J Chem Soc, Faraday Trans 2, 83: 1675
21. Cooper DL, Loades SD, Allan NL (1991) J Mol Struct (THEOCHEM), 229: 189
22. See, for example: Cooper DL, Gerratt J, Raimondi M (1991) Chem Revs 91: 929
23. Cooper DL, Allan NL, Grout PJ (1989) J Chem Soc, Faraday Trans 2, 85: 1519
24. Sag TW, Szekeres G (1964) Math Comput 18: 24553
25. Cioslowski J, Nanayakkara A (1993) J Am Chem Soc 115: 11213
26. Bader RFW (1987) Atoms in a molecule: A quantum theory. Oxford University Press, Oxford
27. Dewar MJS, Rzepa HS (1978) J Am Chem Soc 100: 58
28. Pipek J, Mezey PG (1989) J Chem Phys 90: 4916
29. Cooper DL, Allan NL, Powell RL (1990) J Fluorine Chem 46: 317; 49: 421
30. Derwent RG, Volz-Thomas A, Prather MJ (1989) UNEP/WMO Scientific assessment of stratospheric ozone: Appendix, AFEAS Report, AFEAS, ch. 5, p 123
31. Hampson RF, Kurylo MJ, Sander SP (1989) UNEP/WMO Scientific assessment of stratospheric ozone: Appendix, AFEAS Report, AFEAS, ch 3, p 47
32. Cooper DL, Allan NL, McCulloch A (1990) Atmos Environ 24A: 2417, 2703; Cooper DL, Cunningham TP, Allan NL, McCulloch A (1990) Atmos Environ 26A: 1331
33. Cooper DL, Mort KA, Allan NL, Kinchington D, McGuigan C (1993) J Am Chem Soc 115: 12615
34. McGuigan C, O'Connor TJ, Swords B, Kinchington D (1991) AIDS 5: 1536
35. Kinchington D, McGuigan C (submitted for publication)
36. Cheng L, Tam W, Stevenson SH, Meredith GR, Rikken G, Marder SR (1991) J Phys Chem 95: 10631
37. Matsuzawa N, Dixon DA (1992) Int J Quantum Chem 44: 497
38. Cooper DL, Mort KA, Allan NL, Measures PT, manuscript in preparation

Author Index Volumes 151-173

The volume numbers are printed in italics

Springer-Verlag
and the Environment

We at Springer-Verlag firmly believe that an international science publisher has a special obligation to the environment, and our corporate policies consistently reflect this conviction.

We also expect our business partners – paper mills, printers, packaging manufacturers, etc. – to commit themselves to using environmentally friendly materials and production processes.

The paper in this book is made from low- or no-chlorine pulp and is acid free, in conformance with international standards for paper permanency.